T0254884

Simulation and Statistics with Excel

The use of simulation techniques has increased in importance in recent history, and simulation activities are an important resource for advanced preparation for the organization and execution of events. When formal mathematics is not enough, simulation may be the only option capable of approximating solutions. *Simulation and Statistics with Excel: An Introduction to Business Students* offers a non-rigorous and practical tour of the simulation procedure on computers, using a versatile and accessible resource, the Microsoft Excel spreadsheet. This book covers the concepts essential to understanding the basic principles and approaches of statistical simulation, allowing for the study of complex systems. Aimed at students in business and operational research beginning to use simulation as an instrument for understanding existing or proposed processes, this book will lay solid foundations in understanding simulation experimentation.

Key Features:

- Provides a basis to understand the approaches and principles of simulator experiments.
- Uses a universal and easily accessible resource.
- Introduces simple examples to teach the fundamentals of simulation.

Luis Fernando Ibarra is a Systems Engineer at the Universidad de Los Andes, and an Emeritus Professor at the National Experimental University of Táchira. He obtained his M.Sc. in Industrial Engineering from the Georgia Institute of Technology, and has a diploma in Data Mining from the Central University of Venezuela. In addition, he has an academic management degree (D.M.) in Organizational Leadership from the University of Phoenix.

Simulation and Statistics with Excel
An Introduction to Business Students

Luis Fernando Ibarra

CRC Press
Taylor & Francis Group
Boca Raton London New York

CRC Press is an imprint of the
Taylor & Francis Group, an **informa** business

A CHAPMAN & HALL BOOK

Designed cover image: © Luis Fernando Ibarra

First edition published 2024
by CRC Press
2385 NW Executive Center Drive, Suite 320, Boca Raton FL 33431

and by CRC Press
4 Park Square, Milton Park, Abingdon, Oxon, OX14 4RN

CRC Press is an imprint of Taylor & Francis Group, LLC

© 2024 Luis Fernando Ibarra

ISBN: 978-1-032-69876-2 (hbk)
ISBN: 978-1-032-70154-7 (pbk)
ISBN: 978-1-032-70155-4 (ebk)

DOI: 10.1201/9781032701554

Typeset in Times New Roman
by MPS Limited, Dehradun

DEDICATED TO:

- Luis Fernando Torres Ibarra
- Adrián Alejandro Ibarra Campos
- Máximo Daniel Ibarra Rosales
- Laura Milena Ibarra Vega
- Jorge Enrique Ibarra Vega
- Fernando Javier Ibarra Vega
- Milena Vega Belandria
- Rafaela Ibarra Ortega

Contents

About the Author

Luis Fernando Ibarra

- Systems Engineer, Operations Research mention, Universidad de Los Andes (ULA), Merida, Venezuela.
- M.Sc. in Industrial Engineering, Georgia Institute of Technology (Georgia Tech), Atlanta, USA
- D.M. in Organizational Leadership, with an emphasis on Information Systems and Technologies, University of Phoenix, USA
- Diploma in Data Mining, Central University of Venezuela (UCV)
- Professor of Operations Research and Systems Simulation
- Emeritus Professor at National Experimental University of Táchira (UNET), Venezuela

Foreword

Simulation makes it possible to analyze a system without intervening the system. It is the best-equipped tool for studying complex systems. The construction of simulation models requires the use of computer programming languages. Excel offers an alternative, flexible way to simplify the skills to handle introductory simulations. Its exposition and practical content simplify any formal abstraction to the simulation approach.

This book introduces fundamental concepts to enable the reader to understand and perform simulation experiments using Excel. Simulation is an experimentation tool associated with statistical sampling. The inseparable marriage simulation-statistics requires handling descriptive and inferential statistics to deal with enormous amounts of data. However, the content of this book is designed to facilitate the understanding and use of simulation based on simple chance exercises.

This book describes statistical procedures applied to understand and use the simulation experiment. The proposed exercises, deterministic and stochastic, are didactic examples, suitable to be solved in the spreadsheet environment. They do not correspond to simulations of complex systems, which can be approach using specialized simulation languages. The reader is guided in by using basic models, taking advantage of the facilities offered by a program as widely spread and easily accessible as Excel. Under the Excel platform, the combination of statistics with simulation enhances statistical learning. This interaction puts statistics into action.

The content of this book is useful in an introductory simulation course for students of engineering, management, computing, economics, statistics, and natural sciences. Curricula of studies that include simulation presuppose elementary knowledge of statistics, calculus, and programming. However, even a high school student can benefit from playing numerous games simulated here and learn basic simulation principles.

The first chapter presents the fundamental concepts to understand the approach and basic principles of the simulator experiment. Chapter 2 offers the main functions of Excel for programming many processes and simulation exercises.

Chapters 3 and 4 provide theoretical foundations on random numbers and variables, their generation, and their importance in mimicking the uncertainty inherent in the erratic behavior of most observable systems. It includes procedures expressed in numerical functions to imitate unpredictable system performances.

Chapter 5 formulates the basis for determining the initial conditions and length for the simulation experiment. Simulation length refers to the duration of the simulation, in terms of the sample size of results to be collected to have statistical confidence in the experiment. It includes simple procedures for defining the model run length.

Chapter 6 develops the process of fitting data to various probability distributions. Several goodness-of-fit tests, applied to the most common probability model and their importance to input data to the simulation model, are detailed.

Chapter 7 provides an overview of inferential statistics. Tests used to assess whether a presumption or hypothesis about a set of data can be statistically assumed. The most relevant parametric and nonparametric hypothesis testing models applicable in simulation studies are explained.

Chapter 8 deals with the simulation of several didactic exercises, oriented to strengthen the use of the Excel tool. The examples developed cover knowledge areas, such as games of chance, estimation, inventory control, Markov chains, logistic decisions, sampling, queueing theory, statistics, etc.

Chapter 9 proposes various exercises, aimed at reinforcing the reader's skills in the execution of simulation experiments using the benefits of Excel.

All chapters include practical examples, which provide the necessary tools and learning for the future realization of complex simulations in service systems, manufacturing, gaming, and other knowledge areas.

Fundamentals of the Simulation Experiment

1

1.1 INTRODUCTION

The solidity of any scientific discipline relies on the ability to express its postulates in a precise manner, preferably in quantitative relationships between the causes and effects of the phenomena in its environment. Since every system evolves, periodically evaluating the results of any process is a fundamental requirement to understand and improve its functioning. According to Heraclitus of Ephesus (544 BC): "Everything flows. Everything changes. Nothing remains".

During its existence, a system passes through various possible states. In the case of systems of an unpredictable nature, as time goes by the system develops within its environment facing uncertainties inherent to its existential options. To capture in a quantitative tool the uncertain path of the systems and to determine their evolutionary possibilities, a synthesized digital model constitutes a relevant contribution to evaluate the global performance of processes, besides being in a position to forecast their behavior from the solid perspective of a mathematical computational instrument.

The use of Operations Research methods, specifically the simulation of systems, offers an amazing potential of applicability because this technique is adjustable to any domain that evolves in time. Simulation is a statistical inference tool fundamental for analysis, assessment, and creation in all areas of human knowledge and performance. With the availability of powerful computational tools, any existential or proposed context can be imitated. From celestial phenomena to the quantum behaviors of the particles that compose matter in the micro and mega cosmos can be simulated.

In industrial engineering and scientific management environments, the simulation approach enhances its usefulness in service, manufacturing, material handling, logistics, training, and entertainment contexts. The simulation experiment allows for duplicating the dynamic and uncertain behavior of any production system, including the entire logistics chain, from the arrival of inputs to manufacturing the product and it's positioning in the market.

DOI: 10.1201/9781032701554-1

Computational simulation is an indispensable tool, very effective in studying and solving diverse problems that arise with the advance of universal scientific knowledge. Its flexibility to adapt to existing or proposed environments allows it to be used to verify various scientific hypotheses, becoming an experimental laboratory available to scrutinize new theories and knowledge about the universe. The formal origin of the computational simulation is more than luxurious: the resolution of complex problems in nuclear science for the Manhattan Project during the construction of the first atomic bomb (1942).

The experimental potential of simulation is so immense that no field of human knowledge or natural phenomenon can escape its reach. Simulation is a propulsive tool in the research process. Today and in the years to come, the great challenges of the universe to human understanding will rely on simulation as an essential means of penetration and understanding of our inexplicable existence.

1.2 BASIC CONCEPTS OF SYSTEMS THEORY

A **system** is an entity or set of entities that interact with each other fulfilling an existential purpose. The creator of the General Systems Theory (1947), Ludwig Von Bertalanffy (1901–1972), defines a system as "a complex of interacting elements". That is a set of interrelated elements. The universe, or according to Hawking (2011), the **multiverse**, could be a system composed of multiple subsystems or universes.

Within limited human knowledge, every system is contained within another. Its elements or **entities** are characterized by properties inherent to their condition. These properties are their **attributes**, which differentiate one entity from another. Systems can be **natural** or **artificial** created by man. As time goes by, every system without exception changes its condition or state.

A **state** is a set of characteristics that describe the internal conditions of the system and all its components at any instant in time. The state of a system represents the aspects that completely describe the "position" of the system at any instant in time, and are considered important for studying the future behavior of the system. The state space can be specified in terms of the values of one or more variables, which are called state variables. Starting in a state, every system evolves from one state to another. The actions of the environment on the system (inputs) affect its characteristics and modify its state. That is to say, the inputs or **exogenous variables** affect the system and this is observable in its outputs or results.

A random or **stochastic process** is a system of an unpredictable or random nature whose behavior over time is subject to chance. Generally, it corresponds to a system that can be observed at different instants of time, during which it is only in one of several possible states: E_0, E_1, E_2, …, E_n at respective time instants t_0, t_1, t_2, …, t_n, changing randomly. An **event** is an action that can change the state of the system.

An **activity** corresponds to the occurrence of events at different time instants, from a start event to an end event. A set of **state variables** is used to represent the momentum or state of the system over time. Depending on the variables of interest to the observer, a system can be classified as discrete or continuous. **Discrete variables** are quantitative values that can take on only specific countable values. **Continuous variables** are quantitative values that can take any value within an interval. A system is considered **discrete** if its state variables only change at separate instants of time. The system is **continuous** if its characteristics or states represented by state variables change continuously over time.

1.3 MODELING AND SIMULATION

A **model** is an approximate representation of the structure and relationships between the components of an existing or proposed system, constructed according to an objective and capable of imitating its behavior. The model is expressed through a series of logical relationships and mathematical elements that summarize its performance.

Simulation consists of building a model that mimics the operation of a system or process and experimenting with it over time, to satisfy a certain objective. The objective of a simulation can be to explore, to know, to explain, to predict, to test, to train, to understand, and to entertain.

Computational simulation is the application of digital models for the study and evaluation of processes. Production systems, whether manufacturing or service, correspond to a series of entities that pass through resources where some processing is executed. That is systems where entities, such as materials, parts, people, data, etc., flow in a sequence of steps, occupying various processing resources.

Models can be classified into iconic, analogical, and symbolic. The **iconic model** reflects the structure and properties of the system on a smaller scale, for example, a prototype or a map. **Analogical models** substitute the characteristics of the system by equivalence; for example, a hydraulic system is used in substitution to imitate the flow of vehicles or an economic system. The **symbolic model** imitates the entities of reality and their interactions, using logical-mathematical abstractions. Simulation models are symbolic models.

1.3.1 Building a Model

The construction of a model involves observing the behavior of the system during a certain time, to identify each one of its entities, interactions, and restrictions, and in general, collecting all pertinent information, which allows to representation of the system according to the requirements and objectives of the observer. The construction of the model is the crucial point. Building models is a creative process: a mix of art and science. In model formulation, two potential representations coexist: one referring to the current behavior of the system and the other to its desired behavior. The fundamental principle of scientific research is observation, and its objective is to determine the existential characteristics shown by the systems. Frederick Taylor

(1856–1915), the intellectual father of industrial engineering, recommended in his time studies: "the methodical and detailed observation of the execution of the different tasks by the operators, to rationalize and simplify, to achieve the highest performance".

Any **model** is a simplified representation of the most important aspects of a real system. As such, it is an abstraction that mimics its structure and operation. It will have the detail necessary to satisfy the objective of its formulation. However, it is worth remembering the principle of **Ockham's Razor**: "The simplest model that adequately describes the phenomenon of interest should be used". The model built for simulation becomes an experiment controlled by the observer, subjected to scrutiny, allowing its performance to be analyzed to make inferences about the system it represents. The computational simulation executes the digital model and estimates the system behavior parameters.

1.3.2 Types of Models

Depending on the structure of their state variables, system models can be deterministic or stochastic. The **deterministic model** always shows the same result at equal input conditions. The **stochastic model** contains elements of a random or chance nature that respond in a different and unanticipated way to similar input circumstances.

According to their relationship concerning time, models can be static or dynamic. A model is **static** if its evolution does not involve time management, and its state variables are not influenced by the passage of time. A **dynamic model** mimics the temporal performance of the system so that its state variables change over time.

Mathematical models represent the components and relationship structure of the system through equations and logical relationships. The complexity of some systems requires mathematical models, formulated using differential equations, which reflect the dynamic evolution of their state variables. In addition, complex systems may contain events whose occurrence is a product of chance. This combination may show quantitatively modeled behaviors, whose solution is beyond the reach of formal analytical solution methods. That is, mathematics may be insufficient to deal with systems expressed by endless systems of nonlinear and stochastic differential equations. In such cases, the solution to such complexity must be approximated by simulation experiments.

In fact, for some systems, there is no viable analytical formulation; consequently, the final way to obtain their solution is only using computational experimentation.

In summary, the reason that justifies the use of simulation is to design, understand, or control the operation of complex systems, generally stochastic and dynamic, difficult to manipulate directly. The simulation experiment consists of running the model several times in a computer, recording its results, and if it contains random behavior structures, analyzing those using statistical procedures. Simulation allows for estimating the parameters of the system under study. Parameters can be evaluated in the long or short term. If the estimated value of the parameter depends on the initial conditions of the experiment, the estimation is **transient state**.

It is called **terminal simulation** when specific stopping conditions are defined. On the contrary, if the simulation is performed in the long term and the initial conditions have no effect, it is called **steady-state simulation**, and the parameter estimate is steady state.

1.4 SIMULATION APPROACHES

The main simulation approaches can be classified into the following: discrete-event simulation (DES), continuous-event simulation (CES), system dynamics simulation (SDS), and agent-based simulation (ABS). DES and SDS are traditional modeling techniques, used for the last 60 years. ABS modeling, present since the 1990s, promises to enhance the capacity of the simulator experiment by including the individual behavior of the entities, in addition to their possible collective interactions. However, the Operations Research scientific community and the commercial software available to date, have been slow to assimilate this novel modeling and simulation approach (Siebers 2010). In the ABS approach, "agents are objects with attitudes" (Bradshaw 1997).

In DES, the experiment proceeds in time until any event occurs that changes the state of the system. Time elapses in measured jumps using discrete values. At each instant, every record of the state of the system is updated, and then the system continues to advance until the next event occurs. The simulation of systems whose state variables change continuously is approached using ordinary differential equations or partial differential equations.

DES is suitable for simulating goods or service production systems. These systems are queuing structures or waiting lines, in which a series of customer entities arrive at the service system, where other resource entities can process them during a certain period. The current simulators used in industrial engineering environments, such as Flexsim, Arena, Promodel, Simul8, Simio, etc., are graphical modeling tools that mimic the operation of manufacturing, service, or logistics processes using objects grouped in flow sequences with characteristics and behavior declared by the user.

In the future, it is to be expected that the trio of simulation approaches will merge with the unification of the simulation community's thinking. Consequently, their integration into commercial software that incorporates all three representation strands can be envisioned, targeting simulation on virtual reality platforms. Like the video game industry, simulation will incorporate new computational technologies to achieve realistic simulation models, integrating programs and devices that give the user the feeling of belonging to the simulated environment.

In the case of industrial production, immersive simulation executed in multimedia theaters with human-computer interaction, where protagonist users wearing helmets and data gloves in wireless media, navigate immersed in digital models, validating processes and plant distributions that mimic real environments or proposed 3D manufacturing. Real metaverse production facilities.

1.5 STAGES OF THE DES

Building a model requires knowing the operational details of the system it imitates. It is necessary to determine the components, their interrelationships, and the development environment. The model must offer an adequate representation of the system it imitates, besides including its most important characteristics. The entities to be incorporated in the model depend on the objectives pursued.

A system or process can be synthesized using different models, depending on the scope and objective of the study, and the modeler's preferences. In conclusion, modeling is a highly subjective observational process, as close to art as to science, and every process in the universe can be represented using computational relationships. The construction of a simulation model includes at least the following activities.

1.5.1 Problem Definition

As in any study, it is essential to clearly state the problem of interest, the objectives, and the performance factors to be estimated with the model. The objectives of the simulation study delimit the content and scope of the study. Without clearly defined objectives, modeling and simulation may be useless. The first step in the study of any problem should be to analyze the system to be developed. This includes clearly defining the domain of interest, the objectives, the procedures to be used, and any modifications to be made, review of the structures and behavior of the system. This information can be obtained by inspecting and reviewing specifications, documents, examining the existing system and collecting information from people who are experts in the operation of the processes.

In any simulation study, determining the **objectives** of the experiment is a requirement that precedes the procedures to be performed. Specific objectives can usually be stated and quantified into variables. The development of a simulation plan includes defining various scenarios to be evaluated.

A **scenario** is a specification of controlled inputs arranged as variables in the model to achieve certain responses and judge their behavior. In order not to solve the wrong problem, it is decisive to identify the problem to be solved. The fundamental question is: what are the objectives to be achieved? What results are expected? This requires explicitly stating the purpose of the study, the desired answers, and the problems to be solved.

1.5.2 Observing and Describing the System

The following steps can be used to describe the system:

Steps for describing the system

1. In discrete event simulation, the system is modeled as a flow process of objects, which flow to other resource entities ready to perform service or transformation operations. Pay attention to the system, its environment, and get an overview of the system.
2. Determine the system boundaries and the scope of the study. Is the whole system modeled or only a certain production line? For example, are all customer service activities in a bank simulated, or only the customer service process at the tellers?
3. Draw up a list of all the existing elements in the system, such as machinery, equipment, materials, and any element intended to serve or produce. Determine their relationships, inputs, and outputs.
4. Define the entities that perform processes together with their capabilities.
5. Identify the flowing entities, their flow paths, and their processes. Study existing flow methods and processes.
6. Sampling: data is required to formulate the probability distributions of the events that generate input variables to the system. The data to be collected is oriented by the objectives to be achieved and the answers to be obtained.
   Option 1: Collect between 200 and 300 representative observations for each input variable relevant to the system.
   Option 2: Select a pilot sample and determine the appropriate sampling size. More variation in the real system implies more data to be collected. The results of a simulation are only as good as the data that goes into the model. The motto in computer science "If Garbage In, Garbage Out" is especially true in simulation.

1.5.3 Model Definition

It consists of elaborating a symbolic model that expresses, using mathematical relations, the synthesis of the observation process, including the probability distributions representative of the random events that occur in the system. From the available knowledge about the existing or proposed system, a conceptual model is obtained based on the observed or theoretical knowledge.

A **conceptual model** is the set of assumptions that describe the system entities and interactions it represents, including their limitations, defined by abstractions and simplifications that model the structure and operation of the system analyzed. The conceptual model describes the logical-mathematical image of the reality that is being modeled.

Building a simulation model requires a combination of input data returned by the system and data assumed due to unavailability of values. This logical model is the mathematical and verbal logical representation of the real system, developed through analysis and modeling, symbolized by a flow chart or causal diagram. From this, a mathematical model is derived, composed of expressions, such as equations and constraints, which show the elements and relationships that characterize the system. It should include the elementary components of the system, its domain, variables, parameters, attributes, and their logical relationships, according to the objectives pursued by the project. Subsequently, the **computational model** implements the conceptual model to be executed hundreds or thousands of times on the computer.

1.5.4 Design of Experiment

The design of the experiment indicates how the simulation should be run. Running a simulation involves several repetitions or replicates. Each repetition uses the same data and the same logic, but with different random numbers which produce statistically dissimilar results. A run consists of executing a repetition starting the model in a certain state for a certain length of time or until a stop condition is met.

When designing the experiment, the initial conditions for the execution of the simulation and the duration of each run must be defined. To construct confidence intervals for the random estimates, it is always advisable to run approximately 20 runs or repetitions. That is, several runs to obtain intervals of at least 95% confidence for the estimation of each of the parameters of interest, such as waiting length Lq, number of units in the system L, average time in the system W, average waiting time Wq, probability of leisure Po, output rate, etc.

It is important to remember that a confidence interval can be constructed by calculating the mean and variance of the results of about 20 independent runs. Additionally, the scenarios to be investigated must be specified. For each scenario define the results to be compared. Subsequently, the computational model to be run multiple times on the computer is formulated. Finally, the results obtained from the model should approximate and validate the real observed values of the system.

1.5.5 Model Execution

Use the solution resources to obtain the expected model performance responses. Manage the simulator library to build and run the computer model, which consists of running the computational model using commercially available programs. Selecting and using the right software is critical to simulator success.

1.5.6 Verification and Validation of Results

Verification. It is the process of evaluating the precise fit between the conceptual and computational models to detect conversion errors between both models, to confirm that the computational model corresponds accurately to the conceptual model. Verifying corresponds to checking whether the operational logic of the computational model fits the logic of the conceptual model. That is, if the digital representation imitates the conceptual model satisfactorily. Resuming, validating, and verifying the model consist of ensuring that the model is well built.

Validation. A model is valid if there is statistical confidence in its ability to capture the real process and their results make sense. If data generated by the real system is available, it can be compared with values produced by the simulator model. Various parametric and nonparametric statistical inference techniques are available to evaluate actual responses versus simulated results. These methods are applied to compare the actual observed values of the system with the values generated by the model.

The hypothesis tests defined in Chapter 7 are relevant and applicable to the validation process. One method to validate the data obtained by the simulation with real data is to construct confidence intervals for both data series and compare their similarity (Law & Kelton 1991). In case of not having data on the real process, the validation is at the mercy of the good judgment of the researcher and the knowledge he has of the system.

Finally, for a model to be considered valid, its results must approximate the performance of the system it duplicates. For the results and conclusions to be convincing, the simulation experiment has to acceptably approximate the behavior of the real system.

1.5.7 Sensitivity Analysis

Sensitivity analysis consists of varying the input values of the model and analyzing their impact on its output values. The inquiry "what if?" is the guide. According to the objectives, the relevant parameters and input data of the model are modified, creating different simulated scenarios to observe their impact on the behavior of the system; for example, increasing the arrival rate by 10%, decreasing the service capacity by one or more servers, and analyzing its effect on the system performance factors.

The sensitivity analysis will allow us to evaluate the robustness of the model results to various changes in its input or operating parameters. Since no system remains unchanged, the goal is to take advantage of having full control over the simulator experiment to assess the variability resilience of the results to disturbances in the main input parameters.

To carry out sensitivity analysis, the univariate strategy can be used: change only one value within a range of minimum and maximum values; multivariable: change several values at the same time. Design of experiments (DOE) can be applied to reduce the computational effort caused by the variation of various input factors to the experiment.

It is advisable to assess the impact of changes in the input parameters. If the results are observed to be very sensitive to small changes, it is advisable to verify their origin and veracity.

Finally, the validity of any simulation model is associated with its ability to produce logical, sensible, and credible results. The model has to demonstrate its ability to predict results displayed by the real system.

Introduction to Excel

2

2.1 BASIC DEFINITIONS

The purpose of this introduction is to fix some basic concepts, as well as to review some useful Excel functions and procedures, to increase the understanding of the simulation models to be developed in later chapters. Microsoft Office Professional Plus 2013 version is used.

2.1.1 Cell

An Excel file is called a **workbook** composed of several **spreadsheets**. A sheet may contain separate data or data related to data on other sheets. A **cell** is the basic unit of data storage in the Excel sheet. Cells can store numbers, text, dates, and relationships. They are identified by the intersection between a column and a row of the sheet.

Excel defines the columns by letters: A, B, C, Z, ..., AA, AB, AC, ..., up to the last column referred to as XFD. It counts 16,384 columns. The rows are numbered from row 1 to row 1,048,576. In total, the sheet has 17,179,869,184 cells. Examples A1, B10, and C100 refer to individual cells. **A1** outlines the cell where column **A** and row **1** intersect. Similarly, **B10** indicates the cell located at the intercept of column **B** and row **10**.

2.1.2 Range

It is a set of one or more adjacent cells. The block of cells forms a rectangular shape of cells, which is specified by indicating the two corner cells of the main diagonal separated by two points. The set of cells A1, B1, A2, B2, C1, C2 and D1, D2 in Figure 2.1 can be grouped as the range **A1:D2**. The collection of cells can be edited to perform some operation common to the whole set of cells forming the range. Figure 2.1 shows three rectangular areas or ranges: (1) A1:E3, (2) B5:D15, and (3) F5:F10.

DOI: 10.1201/9781032701554-2

⊿	A	B	C	D	E	F
1						
2		A1:E3				
3						
4						
5						
6						
7						F5:F10
8						
9			B5:D15			
10						
11						
12						
13						
14						
15						

FIGURE 2.1 Three cell ranges.

2.2 RELATIVE, MIXED, OR ABSOLUTE REFERENCE

Excel performs calculation operations using the formulas written in its cells. When copying the contents of a source cell into a target cell, Excel automatically modifies the row and column reference that appears in the source formula. When referencing a cell indicating its position, its column, and row, the column or row may be prefixed with the $ symbol.

If the cell reference is not prefixed with the $ symbol, the cell reference is **relative**. This means when it is moved, Excel automatically adjusts its reference to the next cell. For example, in Figure 2.2, if cell B2 contains the relative reference expression = **A2 + 5;** when the contents of cell B2 are copied into the range of cells in column B, specifically, from B3 to B6, cell B3 receives the formula = A3 + 5; cell B4 is set to = A4 + 5; B5 is left with =A5 + 5; and in B6 appears = A6 + 5. In contrast, if the copy is made on row 2, that is, if cell B2 contains = A2+5, and is copied, for example, into cells C2, D2, and E2, the transformation of the expressions obtained is = B2 + 5, = C2 + 5, and = D2 + 5, respectively.

⊿	A	B	C	D		⊿	A	B	C	D
1		Relative reference	Absolute reference	Mixed reference		1		Relative reference	Absolute reference	Mixed reference
2	1	=A2 + 5	=$A$2 + 5	=$A2 + 5		2	1	6	6	6
3	2	=A3 + 5	=$A$2 + 5	=$A3 + 5		3	2	7	6	7
4	3	=A4 + 5	=$A$2 + 5	=$A4 + 5		4	3	8	6	8
5	4	=A5 + 5	=$A$2 + 5	=$A5 + 5		5	4	9	6	9
6	5	=A6 + 5	=$A$2 + 5	=$A6 + 5		6	5	10	6	10

FIGURE 2.2 Copying in column and moving formulas by column.

An **absolute** or **mixed** reference involves the use of the **$** symbol, which prevents Excel from automatically adjusting rows and columns. For example, in Figure 2.2, cell C2 contains the expression =$A$2+5. When copying the contents of cell C2 into cells C3 to C6, we see that the reference does not change. All cells receive the contents of cell A2+5. That is the expression =$A$2+5 is copied into all the cells involved.

A **fixed mixed** reference combines an absolute column and a relative row, or vice versa, as in $B5 or B$5. In Figure 2.2, the mixed reference =$A2+5 is shown in cell D2. When copied into the range D3:D6, it is noticed that the expression keeps the reference to column A fixed but changes the row from 3 to 6: =$A**3**+5, $A**4**+5, $A**5**+5, and $A**6**+5.

2.3 ADDING AND AVERAGING VALUES IN A RANGE

The function =**SUM**(arguments) sums all values specified as arguments. Each argument can be a range, a cell reference, an array, a constant, or the result of another function. For example, =**SUM**(A1:A10) sums all the values contained in the range of cells A1 through A10.

The function =**AVERAGE**(arguments) returns the arithmetic mean. That is, it adds up all the values that appear in its argument and divides them by the number of values. For example, =**AVERAGE**(A1:A10) returns the average of all the values contained in cells A1:A10.

2.4 CONDITIONAL SUM OF VALUES

The function:

=**SUMIF**(Range of cells; Condition to evaluate).

It executes a conditional sum. It returns the sum of the values that meet a certain criterion or condition. Example:

=**SUMIF**(C5:C100; ">=15").

It sums all values equal to or greater than 15, contained in the range of cells from C5 to C100.

2.5 RESIDUE OF A DIVISION

Occasionally, algorithmic logic requires obtaining the remainder of a division. The function =**MOD**(number; divisor) delivers the remainder of dividing a value by another divisor. For example, =**MOD**(10; 3) yields the value **1**, since 10 divided by 3 is 3 and **1** is left over.

2.6 EVALUATION OF A LOGICAL EXPRESSION

A logical expression or proposition can be True or False. The function:

=**IF**(Expression to be evaluated; Action if Expression is True; Action if Expression is False).

It contains three arguments separated by semicolons, defined to evaluate a logical condition declared in its first argument. If the expression declared as the first argument is True, Excel executes the action indicated by the second argument. If the expression is False, it will result in the action ordered in the third argument.

Example 2.1: Price of a Product According to Demand
The weekly demand for a product is a uniform random variable between 10 and 30 units. If the demand is less than 20 units, the selling price per unit is $100. Otherwise, the unit price decreases by 10%. Simulate the demand for the next 5 weeks.

Solution
To mimic the uniform weekly demand of the product, between 10 and 30 units, cell B2 in Figure 2.3 contains the function =**RANDBETWEEN**(10;30). This function randomly generates an equiprobable integer random value between 10 and 30. Cell C2 contains the expression:

=**IF**(B2< 20;100;90).

	A	B	C	D	E
1	Week	Demand	Price/Unit	Total	
2	1	23	90	2070	→=B2*C2
3	2	30	90	2700	
4	3	22	90	1980	
5	4	17	100	1700	
6	5	24	90	2160	
7					
8		=RANDBETWEEN(10;30)	=IF(B2<20;100;90)		

FIGURE 2.3 Use of the IF function.

It generates a random value in cell B2 and then checks it. If it is less than 20, it returns 100 in cell C2. Otherwise, if it is greater than or equal to 20, it discounts 10% to 100 and arranges in cell C2 the value 90.

From Excel Help: "When a worksheet is recalculated by entering a formula or data in a different cell, or by manually recalculating (press F9), a new random number is generated for any formula that uses the RANDBETWEEN function".

Each arrow shows the expression cell content for first row.

2.7 SEARCHING FOR VALUES

The function =**LOOKUP**() returns a value found in a range of one row or column, matching the position of another value in a second row or column. For example, a range defined over column **A** may contain the codes of the products for sale. If the range declared in column **B** contains the product names, given the code of a product, the function =**LOOKUP**() allows you to find it in column **A**, and at the same time, to find it in column **B**. Automatically modifies the row and column references that appear in the source formula, and indicate the product name. Its syntax is as follows:

=**LOOKUP**(Value to search for; Search range; Result range).

The function takes the value to locate and goes to the search range, where it compares with the reference values until it finds the largest value that overflows the searched value. Finally, from the range of results, it takes and returns the value corresponding to the same position of the search range. The data being looked up must be ordered in ascending way.

Example 2.2: Daily Demand for a Product
The daily demand for a product is random according to the values and probabilities in the following table:

Demand/day	5	10	20	35
Probability	0.30	0.35	0.15	0.2
Cumulative probability	0.30	0.65	0.8	1

Simulate 5 days of demand.

Solution
In Figure 2.4, the range of cells B1:F2 contains the daily demand values and their cumulative probabilities corresponding to each possible demand value. In cell B5, a uniform random value is generated, between 0 and 1, using the =RAND() function, which generates uniform values between 0 and 1. In this case, it is 0.427. In cell

	A	B	C	D	E	F	G	H	I	J
1	Demand/day		5	10	20	35	→ results zone: range C1:F1			
2	Acumulated Probability	0	0.3	0.65	0.8	1	→ searching zone: range B2:F2			
3										
4	Day	Uniform Random Number	Demand							
5	1	0.427	10	→ =LOOKUP(B5;$B$2:$F$2;$C$1:$F$1)						
6	2	0.866	35							
7	3	0.570	10							
8	4	0.119	5							
9	5	0.998	35							
10		↓								
11		=RAND() (value being looked up)								

FIGURE 2.4   Using the =LOOKUP() function.

C5, =**LOOKUP**(B5;$B$2:$F$2;$C$1:$F$1) compares the searched value B5= 0.427, with each value contained in the range B2:F2, until it finds the largest value that does not exceed the searched one. Then, it selects from the range C1:F1 the associated demand value. In Figure 2.4, it is observed that the value 0.427 is between 0.3 and 0.65; therefore, its largest value is 0.65, which results in the associated demand value equal to 10 units. To simulate the remaining days, the range of values A5:C5 must be copied to the range of cells A6:C9. $ symbol keeps the same ranges to look up and results.

2.8 CONDITIONAL CELL COUNTING

The function:

=**COUNTIF**(Range of cells; Condition to evaluate).

It is a conditional counting instruction. It returns a counter which indicates the number of values in the range of cells that meet the condition to evaluate. For example, as shown in Figure 2.5, =**COUNTIF**(C2:C9; "< 20") counts the number of cells in range C2:C9, whose content is less than 20. Since <20 is an expression containing a non-numeric character (<), Excel requires it to be enclosed in quotation marks.

=**COUNTIF**(K2:L50; 20) would count the cells from K2 to L50 whose content is exactly 20.
=**COUNTIF**(A5:A50; "No") returns the number of cells in the range A5:A50, which contain the text **No**.

	A	B	C	D	E	F
1	Day	Uniform Random Number	Demand	Times daily demand < 20		
2	1	0.42	10	5		
3	2	0.35	10			
4	3	0.93	35			
5	4	0.08	5	=COUNTIF(C2:C9;"<20")		
6	5	0.15	5			
7	7	0.74	20			
8	8	0.10	5			
9	9	0.79	20			

FIGURE 2.5   Counting times daily demand is less than 20.

2.9 CONFIDENCE INTERVAL FOR AN AVERAGE

The results of a stochastic simulation estimate values of system parameters. Since these are approximate random values, it is important to delimit a range of values among which it is likely to find the estimated parameter. To do this, lower and upper limits are established around the estimated value, between which, with a certain probability, the value of the estimated parameter is expected to be. The Excel function:

= **CONFIDENCE.NORM** (Alpha; standard deviation; size)

Returns a value that is used to build a confidence interval for a population mean. This function assumes the normality of the population data. Its value is subtracted and added to the estimated mean, allowing us to find lower and upper limits of a confidence interval. The parameters of the confidence function are as follows:

Alpha: Significance level used to calculate the confidence level. The confidence level is equal to $100 * (1 - \alpha)\%$. If $\alpha = 0.05$, it indicates a confidence level of 95% or significance level of 5%.
Standard deviation: Known population standard deviation for the range of data.
Size: Total observations in the sample.

For example, in Figure 2.6, once the average of 13 diameter values under study is estimated as 79.04, we calculate:

=**CONFIDENCE.NORM** (0.05; STDEV.S(B2:B14); 13).

This expression returns a numerical value that represents half of a confidence interval for the average (3.78) (Figure 2.6).

	A	B	C	D	E
1	Wheel	Diameter (mm)	Average	Half of a Confidence Interval	
2	1	88	79.04	3.78	
3	2	79.4			
4	3	87.1			
5	4	77.4	=AVERAGE(B2:B14)		
6	5	83.9			
7	6	68.9	=CONFIDENCE.NORM(0.05;STDEV.S(B2:B14);13)		
8	7	67.7			
9	8	75.7			
10	9	86.1			
11	10	86.5			
12	11	72.9			
13	12	79.7			
14	13	74.2			

FIGURE 2.6 Confidence interval for average.

This value of 3.78 is subtracted from and added to the average to arrive at the lower (79.04 − 3.78) and upper (79.04 + 3.78) limits of the confidence interval. So with 95% confidence, we expect that the value of the diameter is contained between the range [75.26 and 82.82] mm.

2.10 FREQUENCY OF OCCURRENCE OF VALUES

The statistical function: =**FREQUENCY**(range of data; range of classes).

It returns an array containing the frequency for each declared class. This function must be enclosed in square brackets. Close with the keys Ctrl + Shift + Enter pressed at the same time.

Example 2.3: Grouping of 30 Values
Figure 2.7 shows 30 input data arranged in columns A and B. In column D, we arranged the classes to group the data 10 by 10. In column E, from row 2 onward, we order the execution of the function =**FREQUENCY**(A2:B16; D2:D11). Excel returns the count of 5 values less than 10; 4 values less than 20, but greater than 10; etc. However, since it is a matrix function, it must be closed by simultaneously pressing the three keys indicated earlier.

It can be seen that the 30 values are arranged in the range A2:B16. The classes or class intervals are indicated in the range D3:D12. The expression =**FREQUENCY** is declared in the range E3:E12. In cell E3, the expression =**FREQUENCY**(A2:B16; D3:D12) is placed. Subsequently, select the range E3:E12. Then, Ctrl + Shift + Enter.

	A	B	C	D	E	F
1	Values					
2	41	54		Class	Frequency	
3	28	2		10	6	
4	29	10		20	9	
5	12	5		30	2	
6	65	96		40	4	
7	15	100		50	1	
8	88	16		60	1	
9	34	14		70	2	
10	65	94		80	1	
11	33	18		90	1	
12	18	19		100	3	
13	36	5				
14	73	11		=FREQUENCY(A2:B16;D3:D12		
15	9	2				
16	31	18				

FIGURE 2.7 Grouping of 30 values.

Excel performs the count to finally display the frequency distribution. This triple key combination is not necessary in more modern versions of Excel, such as Excel 365.

2.11 DETERMINE IF SEVERAL EXPRESSIONS ARE TRUE

Sometimes it is required to determine if several expressions simultaneously are true. The function:

=**AND**(expression 1; expression 2; expression 3; … ; expression n)

It allows you to determine whether all expressions in a group are true. Excel evaluates each logical expression individually. If all of them are true, the result of the function is true. If only one or more expressions are false, the function returns False as answer.

For example, the expression: =Y(C10=35; D10>= 100; E10 ="High"), simultaneously checks whether: the content of cell C10 is 35; the content of cell D10 is greater than or equal to 100; and if E10 contains the text High. Only if all three expressions are true, the final result of the function is true.

2.12 EVALUATE IF AT LEAST ONE EXPRESSION IS TRUE

As opposed to the function =**AND**(), where all expressions must be true to achieve a resultant True, Excel offers the logical function =**OR**().

The function:

=**OR**(expression 1; expression 2; expression 3; ... ; expression n)

It allows you to determine whether at least one of a group of expressions is true. Excel evaluates each logical expression individually. If one or more expressions are true, the result of the function is true. Only if all expressions are false, the function would return False as a result. For example, the expression =**OR**(C10=35; D10>= 100; G10 ="High"; A10"<=100") evaluates all four expressions. If at least one is true, the result of the function is true.

2.13 ACTIVATING DATA ANALYSIS AND VISUAL BASIC

To activate Data Analysis:

Click FILE / click Options / Ads-Ins / OK

Data Analysis is a module that extends Excel's facilities such as performing regression analysis, constructing histograms, generating random values, running hypothesis tests, etc. It appears in the toolbar when you select the Data option.

Data / Data Analysis

Similarly, if you want to access the Visual Basic program, you need to instruct Excel to activate that option. The sequence of steps given below activates the Visual Basic programming language:

Click **FILE** / click **Options** / select **Analysis Tool Pack-VBA** / **Go** ... / mark **Analysis ToolPack- VBA** / **OK**

2.14 RUNNING MULTIPLE SIMULATIONS

Once the simulation experiment has been run, the result obtained corresponds to a random value or sample of size n = 1. Since the input values are random, it is to be expected that the response will also be random. That is, the model run at each opportunity will deliver different results, although close depending on the level of confidence and accuracy of the simulation length.

To improve the estimation, the model should be run several times. If the sample size or simulation length of the experiment has been previously determined, it may be sufficient to run a total of 20 runs, to finally, from the average of the 20 values obtained estimate the parameter of interest. Although it is an empirical recommendation, the 20 independent results emphasize the length of the simulated sample, and allow the construction of a confidence interval for the main estimated parameter.

Generally users mistakenly settle for the result of a single run of the experiment, unaware that the single value obtained is a random response which becomes an insufficient sample of size n = 1. To correct this, it is advisable to run several independent simulations which guarantee a better approximation to the parameters of interest. To automate the repetition process, it is advantageous to access the Visual Basic programming facility to create a subprogram or macro. A **macro** is a series of actions or instructions that are executed repeatedly and sequentially in an automatic manner. Automating using macro applets requires code programming. However, the Excel toolbar allows the automatic creation of macros. The **DEVELOPER** tab allows access to **Record Macro**. Once activated, this option records all actions that the user executes on the sheet. Although some simulations require writing a bit of instructions, it's a good idea to familiarize yourself with this Excel goodness.

Example 2.4: From a Macro Simulate the Throwing of a Dice
A didactic experiment consists of throwing a balanced dice 10 times and determining its average value. Moreover, we want to define a macro to perform 10 independent simulations.

Solution
Figure 2.8, row 1, shows the column headings. A1: Throw; B1: Result; D1: Simulation; E1: Average. In cells A2–A11 and D2–D11, the sequences numbering the order of the

	A	B	C	D	E	F	G	H
1	Throw	Result		Simulation	Average			
2	1	4		1	3.4			
3	2	1		2	3.8		Click to realize	
4	3	2		3	4.1		10 simulations	
5	4	6		4	3.8		10 Throws each	
6	5	4		5	3.3			
7	6	5		6	4.6			
8	7	5		7	3.1			
9	8	1		8	2.7			
10	9	4		9	3.7			
11	10	3		10	3.1			
12	Average for 10 Throws:	3.5		Average for 10 simulations:	3.56	=AVERAGE(E2:E11)		
13 / 14	=AVERAGE(B2:B11)		=RANDBETWEEN(1; 6)					
15 / 16	Estimated half CI:	0.343	→	=CONFIDENCE.NORM(0,05; STDEV(E2:E11); 10)				
17	Confidence Interval for Average:	3.217	3.903	=E12-B15 / =E12+B15				

FIGURE 2.8 Throwing of a dice: 10 repetitions with Excel macro.

throws are entered. To obtain the result when the dice is thrown, in Figure 2.8, cell B2 the function =AVERAGE(1; 6) is defined, and copied to cell B11. By means of =AVERAGE(B2:B11), in cell B12, the average of the 10 throws is calculated.

It is observed that the answer of interest appears in cell B12. In addition, 10 independent repetitions are desired, the results of which will be displayed in column E, specifically, in the range E2:E11.

2.15 MACRO CREATION PROCEDURE

The procedure for programming a macro that simulates the throwing of a dice, and performs several repetitions, is summarized in Table 2.1:

TABLE 2.1   Procedure for programming a macro

1. In the ribbon, select **VIEW / Macros / View Macros**
2. Type the name of the macro: **Dice**
3. Click on the **Create** button
4. Position the cursor between **Sub Dice**() and **End Sub**. Type the following instructions:

Range("E2:E11").ClearContents
n = Val(InputBox("How many experiments to perform? N= ", "Throwing", 10))
For i = 1 To n
    Cells(i + 1, "E") = Range("B12").Value
Next i

2.16 EXECUTING A SUBPROGRAM

To execute the macro called **Dice**, click on the ribbon:

Run / Run Sub / UserForm F5 / OK

During the execution and from the worksheet, the macro deletes all contents of the range E2:E11. Then it requests the number of simulations to be performed, whose value is stored in the variable **n**, and returns to the macro work area. Then it executes the instructions included in **For** and **Next** repeatedly **n** times. It starts with the value i = 1, then i = 2, 3, …, **n** incrementing one by one, until the last value of **n** is reached. In this case, the only internal instruction **For/Next** will take the value contained in cell **B12**, and write it to cell **i+1** of column E, varying i from **1** to **10**. Excel fills a column vector of 10 values which will be stored in the range **E2:E11**.

To return to the Excel sheet and observe results, you can click on the Excel icon at the top of the sheet. It will show the 10 values corresponding to the independent runs. Subsequently, a better estimate of the parameter can be obtained by repeating and averaging the **n** (10) averages in cell **E12**. It means averaging 10 averages.

Figure 2.8 includes the calculation of a 95% confidence interval for the average, as well as a form button called **Launch** which automates the macro execution procedure, the construction of which is presented in Section 2.18.

2.17 STORING A MACRO

To save the macro, you must select **FILE / Save As /** give file name **/ Excel Macro-Enable Workbook / Save**.

The Excel sheet containing macro code will be saved in a macro-enabled .xlsm file.

2.18 AUTOMATING WITH FORM CONTROLS

To automate the macro execution procedure, you can insert Form Controls. Form Controls are program-user interaction dialog formats. From Excel toolbar ribbon select **DEVELOPER**. If it does not appear it could be activated by the following procedure:

1. **FILE / Options / Add–Ins / Analysis ToolPak-VBA / Go /** mark **Analysis ToolPak – VBA / OK**
2. **FILE / Options / Customer Ribbons /** mark **Developer / OK**.

Insert option appears on the **DEVELOPER** ribbon. In its selection list, the available Form Controls are displayed. The procedure to follow is:

DEVELOPER / Insert / Button (Form Control).

The cursor turns into a small cross. Position and click the cross cursor in the free area of the worksheet. Select the name of the macro associated with this button, in this case **Dice**, and finally click **OK**.

You can right click the button to assign a text name to it. Type the name that identifies the button and click outside the button. Click the button each time you want to execute the repetitions, even for values of **n** less or greater than 10.

2.19 CONSTRUCTION OF A HISTOGRAM: EXPONENTIAL DISTRIBUTION

The lifetime of a battery is randomly exponentially distributed, with an average of 20 months. Simulate and construct a histogram for the lifetime of 1,000 batteries.

Solution
See procedure in Table 2.2. The lifetimes of a thousand batteries will be generated, arranged in column B, range B2:B1001 (Figure 2.9).

TABLE 2.2 Histogram of an exponential distribution with average 20

1. Write the exponential generator expression in cell **B2: = – 20*LN(RAND())**
2. Copy cell **B2** into cells **B3 to B1001**.
3. In the ribbon toolbar Click **DATA**
4. Click **Data Analysis**
5. Select **Histogram** and click on **OK**
6. In Input range enter **B2:B1001**
7. Under Output options select **Output range**. Place the cursor and indicate a cell for the output, for example, D1.
8. Check the **Create graph** option
9. **Accept**

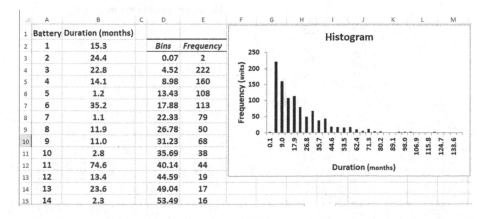

	A	B	C	D	E	F ... M
1	Battery	Duration (months)				Histogram
2	1	15.3		*Bins*	*Frequency*	
3	2	24.4		0.07	2	
4	3	22.8		4.52	222	
5	4	14.1		8.98	160	
6	5	1.2		13.43	108	
7	6	35.2		17.88	113	
8	7	1.1		22.33	79	
9	8	11.9		26.78	50	
10	9	11.0		31.23	68	
11	10	2.8		35.69	38	
12	11	74.6		40.14	44	
13	12	13.4		44.59	19	
14	13	23.6		49.04	17	
15	14	2.3		53.49	16	

FIGURE 2.9 Partial data for duration of 1,000 batteries.

2.20 NORMAL GRAPH USING THE HISTOGRAM FUNCTION

The weight of an individual can be approximated to a normal distribution with mean of 75 kg and standard deviation of 15 kg. Simulate the weight of a thousand people and plot those values in a histogram.

Solution

After filling in column **A** with sequential numbers for each person (Figure 2.10), place in column **B** the random generator used for a normal distribution with mean of 75 in cell **B2**:

=**NORM.INV**(RAND(); 75; 15)

To get Figure 2.10, follow the sequence of steps 2–9 of the procedure described in Table 2.2.

An alternative to obtain a frequency distribution is using the Excel function =**Frequency** (range of data; range of classes). Now we are going to make a new histogram directly by using option Excel included in Data Analysis. It is necessary to indicate its two arguments: the range containing the random data and the range of classes.

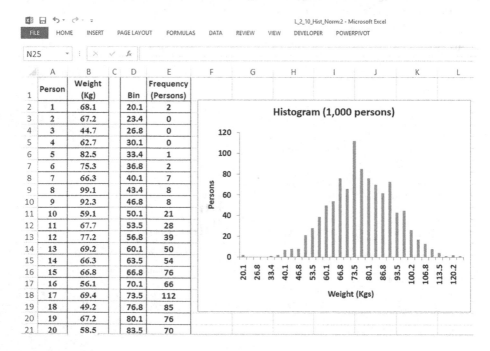

FIGURE 2.10   Histogram: weights of 1,000 people.

TABLE 2.3   Plotting normal distributions with the Histogram option

1. Generate one thousand normal values in column B, range B2:B1001. To do this, the function =ROUND(**NORM.INV**(RAND();75;15);1) is placed in cell B2.
2. Copy cell B2 in range B3:B1001
3. In column C, cell C2, enter 10
4. In cell C3 enter =C2+5
5. Copy cell C2 into cell range C3:C28
6. Select the empty range D2:D28. To do this, hold down the left mouse button from D2 to D28, and release the cursor. The selected area becomes blue and selected.
7. Type =**FREQUENCY**(B2:B1001; C2:C28)
8. Simultaneously press the keys **Ctrl + Shift + Enter**

For this example, newly declare normal data with mean of 75 and standard deviation of 15 kg. Random data are in the range B2:B1001 and their classes will be taken of amplitude 5, and arranged in the range C2:C28. The steps to be followed are shown in Table 2.3.

Excel outputs the frequency or number of values contained in each class or bin. Figure 2.11 presents only the first 17 individuals and their weight. However, the frequency distribution shown corresponds to the weights grouped for 1,000 individuals.

	A	B	C	D	E	F	G	
1	Individual	Weight (kgs)	Bins (kgs)	Frequency				
2	1	94.9	10	0				
3	2	69.1	15	0				
4	3	57.3	20	0	{=FREQUENCY(B2:B1001;C2:C28)}			
5	4	73.9	25	1				
6	5	63.7	30	2				
7	6	73.1	35	4	ROUND(NORM.INV(RAND();75;15);1)			
8	7	77.4	40	6				
9	8	115.4	45	14				
10	9	63.4	50	22				
11	10	81.7	55	39				
12	11	68.2	60	62				
13	12	55.3	65	97				
14	13	70.5	70	114				
15	14	84.4	75	129				
16	15	97.4	80	136				
17	16	94.6	85	116				
18	17	91.2	90	90				

FIGURE 2.11   Distribution of the weight (thousand observations).

2.21 RANDOM GENERATION WITH DATA ANALYSIS

To generate random values of some probability distributions, Excel offers an add-in that can be installed beforehand. By selecting **DATA/Data Analysis/RAND Number Generation**, the window shown in Figure 2.12 is obtained.

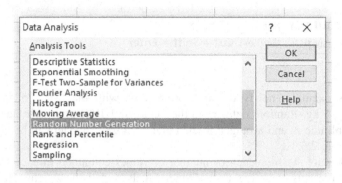

FIGURE 2.12   Generation with data analysis.

The random values are arranged by column. The information to be entered includes the following:

- **Number of variables**: Number of columns or variables to be generated.
- **Number of random numbers**: Number of random values to fill for each column.
- **Distribution**: Select the type of probability distribution of the values to be generated.
- **Parameters**: Indicate the parameters of the distribution.
- **Output options**: Choose the output area to receive the answer.

Example 2.5: Generating Normal Random Numbers
Use Excel Data Analysis to generate 20 random values, in two columns, from a normal distribution of mean 100 and deviation 20 (Figure 2.13).

2.22 CALCULATING PROBABILITIES IN EXCEL

Excel offers several probability functions designed to simplify the task of calculating probabilities and their use in the simulation experiment. Their syntax and usage is summarized below.

	A	B	C	D	E	F	G
1			Random Number Generation			?	X
2	89.24	96.92	Number of Variables:		2		OK
3	102.75	143.00	Number of Random Numbers:		10		Cancel
4	106.03	109.70	Distribution:	Normal			Help
5	120.93	126.04	Parameters				
6	106.53	112.88	Mean =		100		
7	88.85	87.30					
8	79.96	99.65	Standard deviation =		20		
9	100.11	110.50					
10	77.27	82.04	Random Seed:				
11	88.51	118.05	Output options				
12			⦿ Output Range:		SAS2		
13			○ New Worksheet Ply:				
14			○ New Workbook				

FIGURE 2.13   Normal sample using data analysis.

The last argument of the probability functions corresponds to whether or not the cumulative probability calculation is desired. If x is a random variable represented by a probability density function f(x), then its cumulative probability function F(x) measures the probability that the variable x takes any value below a reference value X. If the argument **cumulative** is 1, Excel will return the cumulative probability for the range of values below the indicated X. If the value is zero, it will return only the probability of the requested value X.

2.22.1 Binomial Distribution

=**BINOM.DIST** (Successes; Trials; Probability of success; cumulative up to X).

The argument Cumulative can be false (0), or true (1). If it is 1, it returns the cumulative function F(x).

Example 2.6: A quality inspector checks boxes containing 20 units. The probability of randomly selecting a defective unit is 0.3. Determine the probability that the inspector finds exactly three defective units.

Solution
To calculate this probability use the function:

=**BINOM.DIST**(3; 20; 0.3; 0)= 0.0716.

It is concluded that the probability that the inspector finds exactly three defective units is 0.0716.

2.22.2 Poisson Distribution

=POISSON.DIST(X events occur; Poisson average; Cumulative)

Example 2.7: The number of defects in a 100-m roll of cloth is a Poisson random variable of mean 2.5.

A.   What is the probability that a roll has no defects?

=POISSON.DIST(0; 2.5; 0) = 0.082085

The probability that a roll has no defects is 0.082085.

B.   Has 2 or more defects? =1- POISSON.DIST(1; 2.5; 1) = 0.7127025.

The probability that a roll has 2 or more defects is 0.7127025.

2.22.3 Exponential Distribution

=EXPON.DIST(X; 1/exponential average; Cumulative)

X is the value of the variable to be evaluated. 1/exponential average corresponds to λ, average of events that occurs during a unit time.

Example 2.8: The time T taken by a worker to prepare a machine is exponential averaging 25 minutes. What is the probability that the worker takes less than 25 minutes to prepare the machine?

Probability(T $\leq$ 25 min.) =EXPON.DIST(25; 1/25; 1) = 0.6321

The probability that a worker takes less than 25 minutes is 0.6321.

2.22.4 Hypergeometric Distribution

=HYPGEOM.DIST(ss; n; sp; N; cumulative)

ss: Number of successful elements expected to be found in the sample selected
n: Size of the sample or number of elements selected
sp: Number of successful elements in the population
N: Population size

Last argument: If it refers to the cumulative probability distribution, accumulates is 1, if not 0.

Example 2.9: A deck of Poker contains 52 cards. Two cards are drawn at the same time. What is the probability of getting an Ace?

Hits in the sample ss = 1. Sample size n = 2, Hits in the population sp = 4, population size N = 52.

=HYPGEOM.DIST(1; 2; 4; 52; 0) = 0.145

The probability that the player receives an Ace is 0.145. That is, a Poker player who initiates reception, can expect to receive an Ace approximately every 7 deals. (1 / 0.145 = 7).

2.22.5 Lognormal Distribution

=LOGNORM.DIST(variable x; Average; Deviation; Cumulative)

Example 2.10: The annual income of the inhabitants of a municipality follows a lognormal distribution of parameters mean 10 and variance 1.5. What percentage of individuals is expected to have an income of less than $100,000?

=LOGNORM.DIST(100000; 10; 1.5; 1) = 0.843.

84.3% of the individuals have incomes below $100,000 per year.

2.22.6 Weibull Distribution

=WEIBULL.DIST(x; shape; scale)

Example 2.11: The failure time of a flat panel display is random Weibull with shape α = 1/3 hours and scale β = 200. Estimate the probability that a display will last more than 2,000 hours.

P(x > 2000 hours) = 1 − **WEIBULL.DIST**(2000;1/3; 200;1) = 0.116

2.22.7 Gamma Distribution

=GAMMA.DIST(x; Alpha shape; Beta scale; cumulative)

Example 2.12: The length of time in years of a brand of car is a Gamma random variable of parameter form 1.5 and scale 4.7. What is the probability that a car will exceed 5 years?

=1− GAMMA.DIST(5; 1.5; 4.7 ;1) = 0.546

54.6% of the cars exceed 5 years.

2.22.8 Normal Distribution

=NORM.DIST(X; Mean; Deviation; cumulative)

Example 2.13: The weight of the contents of mayonnaise containers varies randomly according to a normal distribution, with mean 900 grams and standard deviation of 35 grams.

    A. What is the probability that a container weighs less than 850 grams?

=NORM.DIST(850; 900; 35; 1) = 0.076.

The probability that a container weighs less than 850 grams is 0.076.

    B. What percentage of containers weigh between 850 and 950 grams?

=NORM.DIST(950; 900; 35; 1) − NORM.DIST(850; 900; 35; 1) = 0.84687.

That is, 84.69% of the containers are expected to contain between 850 and 950 grams of mayonnaise.

2.22.9 Beta Distribution

=BETA.DIST(X; Alpha; Alpha; Beta; Cumulative; [A]; [B])

Example 2.14: The daily percentage of defective items produced in a factory is represented by a Beta distribution of parameters Alpha = 2.5 and Beta = 65.

    A. What is the probability that in a day, the percentage of defective items is less than 2%?

=BETA.DIST(0.02; 2.5; 65; 1) = 0.2472.

The probability that in one day the percentage of defectives is less than 2% is 0.2472.

    B. During a year, how many days are expected to have more than 3% defective items?

Probability of getting more than 3%: =1− BETA.DIST(0.03; 2.5; 65; 1) = 0.5487.
The expected number of days with more than 3% defective = 0.5487 ∗ 365 = 200.3 days.

2.23 FUNCTIONS FOR RANDOM GENERATION IN EXCEL 2010

Excel has updated the names of its statistical functions. All functions whose name include the word **INV** correspond to random value generators. The following expressions are useful for modeling applications under simulation in Excel. Each name suggests the probability distribution it represents. Classical random generation procedures will be shown later in Chapter 3.

Uniform(0,1): =RAND()
Uniform(a,b): =RANDBETWEEN(a;b) uniform discrete values between a and b.
Uniform(a,b): =a+(b-a)∗RAND() continue values between a and b.
Geometric: =ROUNDDOWN(LN(RAND()) / LN(1-p); 0)
Poisson: =BINOM.INV(average Poisson λ / 0.001; 0.001; RAND())
Exponential: =−1/ λ∗LN(RAND())
Binomial: =BINOM.INV(trials n; probability of success p; RAND())
Normal(0,1): = NORM.S.INV(RAND())
Normal(μ;σ): = NORM.INV(RAND(); μ; σ)
Lognormal: =LOGNORM.INV(RAND(); μ; σ) μ and σ are mean & standard deviation of ln(x)
Weibull: =α ∗(-LN(RAND()))^(1/ β) where shape β, and scale α
Gamma: =GAMMA.INV(RAND(); α; β) where shape α; scale β
Beta: =BETA.INV(RAND(); α; β; a; b)
Triangular: = c + (a + RAND() ∗ (b-a) - c) ∗ SQRT(RAND())
Gumbel or Extreme Value: μ − β ∗ ln(−ln(RAND()))

Uncertainty and Random Numbers

3

3.1 IMITATION OF UNCERTAINTY

The events that occur in a system can be governed by chance, so that, although their possible responses are known, the output at a given instant is not known with certainty. Processes governed by chance are analyzed in engineering by evaluating the degree of variability of their behavior using probability models. Probability measures the degree of uncertainty of occurrence of those events whose result or output cannot be anticipated with certainty.

The **sample space** is the set of different values that an event can show. For example, the number of customers at the door of a bank at opening time is variable and cannot be known in advance. The volume of customers entering at the opening of the door of a bank agency on Monday could be 46 people, on Tuesday 62, on Wednesday 54, and so on. In other words, an uncertain variable whose value is impossible to anticipate. Nor is it possible to know with certainty how long it will take a teller to attend a customer's transaction. Even less, excluding quantum physicists and others, it is impossible to successfully predict the side that will appear when throwing a balanced die or a coin.

After observing the evolution of an event over time, and recording the number of times an outcome occurs, it is possible to infer the proportion of times it could happen. This proportion of occurrences, according to the **strong law of large numbers**, corresponds to the probability calculated from the frequency of the event. If an independent number of **n** casual observations is obtained to estimate the average of a population, the sample average tends to μ as **n** tends to infinity. That is to say, the average of **n** values of the variable **x** will get closer to the population average as more values of the variable are taken.

The creation of mathematical models requires the use of symbols that represent the events and behavior of the system or process under study. The characteristics or properties of a system are represented by variables. These record some particularity of interest, which fluctuates between different values. A **variable** is considered **deterministic** if the value it will assume at a given instant is known with certainty. An **unsystematic variable** takes random values within a certain range of feasible values. However, with certainty, it is not possible to anticipate the value it will show. Given such

32

DOI: 10.1201/9781032701554-3

uncertainty, it is only possible to associate a possibility or probability that the variable will take any of its possible values.

In addition, a **variable** is **discrete** if its range or range of values is finite. For example, number of customers waiting to be served: 0, 1, 2, 3, 3, 4, ..., n. A **variable** is **continuous** if it can take infinite values between any two of its values: the length of a table, the height or weight of a person, are continuous variables because between any pair of values, for example, between 100.5 and 100.6 cm there are infinite values of length (100.52 cm, 100.734 cm, etc.). If a variable is not affected by events occurring within the system, it is called **exogenous** or external. If, on the contrary, its value is affected by eventualities within the system, it is referred to as **endogenous** or internal.

3.2 RANDOM NUMBERS

Since the behavior of many systems is influenced by various uncertain factors, obtaining random numbers is fundamental to the process of simulating unpredictable events. Random inputs to the model are symbolized by variables generated from probability distributions.

To mimic the uncertainty or randomness of stochastic events, algorithmic procedures based on sequences of random numbers are used. The instant of time at which entities arrive at a process, and the quantity that arrives, are uncertain, and can be expressed by random variables. The service time required to serve an entity also varies from one entity to another. Systems subject to uncertainty are called **stochastic** or **random systems**.

Random numbers correspond to series of numerical values obtained by procedures that mimic chance. They are numbers drawn independently with equal probability from the population 0, 1, 2, 3, ..., 9. The first methods for obtaining unplanned values were manual: drawing cards and throwing dice.

If 10 identical cards numbered from 0 to 9 are placed in a bag, the procedure of drawing a card at random and recording its result is a purely random generation procedure; since it is not possible to anticipate the value that will be obtained at each drawing. In this case, the value obtained at each opportunity may correspond to a random digit.

If the process is repeated a certain number **n** of times, returning the extracted token to the bag, a pure chance sequence of length **n** is obtained.

For example, if **n** = 3 for 6 sequences results like the following could be achieved (Table 3.1):

TABLE 3.1 Six random sequences each of three extractions

SEQUENCE	1	2	3	4	5	6
Random Tokens	3; 0; 7	5; 2; 9	0; 5; 5	3; 8; 8	2; 4; 1	1; 6; 7

A sequence corresponds to three draws with replacement. In a draw with replacement, each of the 10 numbered tiles has the same chance of being selected. That is, its chance of being selected is one in ten (1/10).

3.3 PURE RANDOM NUMBERS

Random values achieved by procedures based on physical processes, such as a roulette wheel, atmospheric noises, electrical pulses, extractions, a bag and 10 chips, etc., are called pure random numbers. In such processes, it is usually impossible to manipulate the unpredictable nature of the phenomenon and to anticipate its output.

Nowadays, to obtain pure random numbers, random event generating machines are available: devices connected to a computer, capable of producing sequences of numbers governed by chance. These machines use physical phenomena such as unplanned source coming from the decay of radioactive materials or from electrical noise from a semiconductor diode.

The electronic circuitry of the generator through its serial port converts the noise into bits; it groups them into bytes to be arranged in the computer, guaranteeing their unsystematic purity. Cryptographic systems require millions of pure accidental values to generate access keys and perform various operations that prevent invisible threats to computer systems from attackers and spies.

Commercial devices such as the Whitewood Entropy Engine employ quantum random processes of atoms and photons to generate pure chance numbers at the rate of hundreds of Mbps (350 million of bits per second), transmitted directly to any computer and its applications. If a hacker wanted to manipulate the values generated by these devices and force them to generate deterministic numbers, it would require calculating the movements of thousands of electrons in the resistor or semiconductor, which to date does not seem feasible.

3.4 PSEUDO-RANDOM NUMBERS

Computers are logical and deterministic machines. The algorithms used by computing machines are based on recursive algorithms generated by mathematical formulas, whose results are completely predictable. Their recursion consists of having an initial value X_o (seed), and then, starting from that first value, substituting it in the formula of the corresponding algorithm. In this way, the series of random values X_1, X_2, X_3, ..., X_n is obtained. The values thus obtained are not pure unplanned. However, well designed, the algorithms deliver statistically acceptable values called pseudo-random numbers.

These numbers are generated from deterministic procedures although they appear random. If you know the algorithm, you can obviously anticipate the series of numbers it will produce. Although they are predictable numbers, their methods are subjected to severe scrutiny of statistical reliability, so that, henceforth, they will be called random numbers, obviating their real pseudo-random condition. That is, they are false random values although if they meet a series of requirements and pass a series of tests, they are statistically admissible as if they were governed by chance.

The statistical fitting software EasyFit handles the widely used chance generation algorithm called Mersenne Twister, which produces a repetition cycle with period $2^{19937}-1 > 10^{6000}$. This method passes Diehard's battery of statistical quality tests: a powerful set of 18 statistical tests, developed by mathematician George Marsaglia, aimed at investigating the quality of any casuistic generation algorithm. This collection of tests evaluates the generated unplanned numbers, looking for correlations, trends, deviations, or periodic repetitions.

The uncertain behavior of the components of a system can be represented by probability distributions. Probability distributions are suitable for predicting the performance of random variables. A value obtained from a probability distribution is called a random variate. Most programming languages and simulator packages include generators capable of producing random variate sequences acceptable to simulation models. For the purposes of academic experimentation, commercially available pseudo-random generators are sufficient. However, the use of random values from poorly designed algorithms invalidates the simulation results. For illustrative purposes only, one of the countless iterative algorithms available albeit obsolete used to obtain random values, known as the central square method, is shown below.

3.5 CENTRAL SQUARE METHOD

It consists of taking an initial number X_0, of several digits, squaring it, and selecting several digits from the center of the number to form the first value X_1. Then iteratively repeat the procedure. To express the number between zero and one, each sequence of digits generated is prefixed with a zero.

Example 3.1: Central Square Method

Let $X_0 = 3567$. Use the central square method to generate four random four-digit numbers (Table 3.2).

TABLE 3.2 Application of the central square method

I	X_I	X_i^2	RANDOM NUMBER
0	3567	12723489	0.7234
1	7234	52330756	0.3307
2	3307	10936249	0.9362
3	9362	87647044	0.6470

3.6 LINEAR CONGRUENTIAL METHOD

Linear congruential random number generators are recursive algorithms having the form:

$$Z_{i+1} = (a * Z_i + c) \bmod m$$
$$R_{i+1} = Z_{i+1}/m$$

where **a**, **c**, **m**, and Z_i are non-negative integer values. The value **m** is the modulus degree; and Z_0 is the seed or initial value. For example, if $a = 7$, $c = 2$, $m = 45$, and $Z_0 = 3$, generate 3 random values using the linear congruential method (Table 3.3).

TABLE 3.3 Random generation: linear congruential method

$Z_i = (7 * 3 + 2) \bmod 45$	Random values R_i
$Z_1 = 23 \bmod 45 = 23$	$R_1 = 23/45 = 0.511$
$Z_2 = (7 * 23 + 2) \bmod 45 = 28$	$R_2 = 28/45 = 0.622$
$Z_3 = (7 * 28 + 2) \bmod 45 = 18$	$R_3 = 18/45 = 0.400$

Solution
The three Z_i values obtained are: 23, 28, and 18. The linear congruential method gives a sequence of random values $0 \leq R_i < m$. The quotient $\frac{R_i}{m}$ transforms the values within the range $0 \leq R_i < 1$. By expressing the previously obtained Zi values in the decimal range, after dividing each value by 45, the random values R_i are obtained.

Figure 3.1 gives a sample of five random numbers. When creating more values, the method will keep repeating the same sequence. Its cycle is only nine different values. The simulation of a complex stochastic system requires millions of random values, so the parameters of the congruential method must be improved so that the

	A	B	C	D	E	F
1	m	a	c	Zo		
2	45	7	2	3		
3						
4						
5	i	a * z(i-1) + c	Z(i)	R(i) = Z(i)/m		
6	1	23	23	0.511		=C6/$A$2
7	2	163	28	0.622		
8	3	198	18	0.400		
9	4	128	38	0.844		
10	5	268	43	0.956		
11						
12		=$B$2*C7+$C$2	=MOD(B8;$A$2)			

FIGURE 3.1 Congruential generation.

method achieves sufficiently long generation cycles. Generally, the modulus **m** is defined as the power of 2 raised to the number of bits of the computer word used. With **m** = 2^{32}–1, an acceptable cycle of more than 2 billion different values is guaranteed.

3.7 RANDOM NUMBER GENERATOR FUNCTIONS

Generating uniform random numbers between 0 and 1 is fundamental to the simulation process, since from uniform random digits, values of random variables are derived to model events fitted to any probability distribution. Calculators and computers have automatic functions for this purpose. Some pocket calculators incorporate the **Rnd#** function. When this key is pressed, a uniform random value of at least three digits is displayed on the screen. Each random value has a probability of occurrence of $1/10^{\text{digits}}$.

The Excel program offers the function =**RAND**(), which generates uniform random values with 15 decimal digits. That is, each value between 0 and 1 has an equal probability of occurring.

3.8 SIMULATION OF A COIN FLIP

When a balanced coin is tossed on a firm surface, only one of two possible outcomes is obtained: "Head" or "Tail". You can mimic the procedure of tossing a coin using the Rnd# function of a pocket calculator or the =**RAND**() function of Excel.

Since the **RAND**() function generates values with equal probability of occurrence between 0 and 1, all its values are equally assigned between the two possible outcomes, so that half of the values generated by the =RAND() function, from 0.000 to 0.500 correspond to the outcome "Head"; and the remaining values, from 0.500 to 0.999, to the outcome "Tail". Table 3.4 shows a possible sequence of values obtained by manually simulating five throws:

TABLE 3.4   Simulation of five tosses of a coin.

FLIP	RANDOM NUMBER **R**	COIN RESULT
1	0.432	Head
2	0.951	Tail
3	0.072	Head
4	0.350	Head
5	0.653	Tail

	A	B	C	D	E	F
1	Flip	Random	Coin			
2	1	0.103	Head		→=IF(B2<0.5; "Head";"Tail")	
3	2	0.448	Head			
4	3	0.523	Tail			
5	4	0.043	Head			
6	5	0.808	Tail			
7						
8		=RAND()				

FIGURE 3.2 Excel simulation of five coin tosses.

Figure 3.2 shows simulation results in Excel when a balanced coin is tossed. The decision function **=IF()** allows to evaluate the random value obtained and to decide "Head" when **=RAND()** is between 0.000 and 0.4999. The remaining values from 0.500 to 0.999 correspond to the event "Tail".

The decision criterion "Head" or "Tail" can be reversed. For example, if the RAND () is less than 0.5, the decision can be "Tail"; and if greater than 0.5 then "Head". What matters is to assign the equivalent fraction of RAND() values to each possible outcome.

3.9 SIMULATION OF DICE THROWING

Different procedures can be defined to imitate the throwing of a dice. Among others, drawing with replenishment of a container with six similar tokens, previously numbered from 1 to 6, so that each token imitates a result on the side of a dice. A throw corresponds to drawing a token, recording its numerical value as the value obtained by rolling a balanced dice and returning the token to the container. In this case, the value marked on each token mimics the value corresponding to the top side of a falling dice.

To simulate the throwing of one or more dice, in Excel, you can use the function:

=RANDBETWEEN(minimum value; maximum value).

This function delivers equiprobable random values, between the minimum and maximum values. In the case of =RANDBETWEEN(1;6), it distributes all the RAND () values among the six results in equal parts. Figure 3.3 shows the Excel output when mimicking the roll of a dice (Table 3.5):

	A	B			A	B
1	Throw	Dice		1	Throw	Dice
2	1	3		2	1	=RANDBETWEEN(1;6)
3	2	5		3	=A2+1	=RANDBETWEEN(1;6)
4	3	3		4	=A3+1	=RANDBETWEEN(1;6)
5	4	4		5	=A4+1	=RANDBETWEEN(1;6)
6	5	5		6	=A5+1	=RANDBETWEEN(1;6)

FIGURE 3.3 Simulation of a dice roll.

TABLE 3.5 Manual simulation: five bets

ROLL	JOHN'S DIE 1	JOHN'S DIE 2	SUM OF DICE: JOHN	PETER'S DIE 1	PETER'S DIE 2	SUM OF DICE: PETER	JOHN'S BALANCE $S_J(N)$ [$]	PETER'S BALANCE $S_P(N)$ [$]
Start	-	-	-	-	-	-	100	100
1	3	6	9	5	6	11	95	105
2	1	6	7	4	5	9	90	110
3	6	4	10	5	5	10	90	110
4	3	6	9	4	2	6	95	105
5	2	3	5	3	3	6	90	110

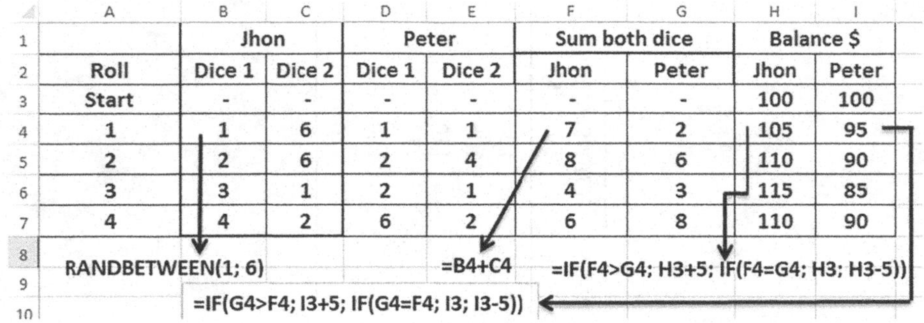

	A	B	C	D	E	F	G	H	I
1		Jhon		Peter		Sum both dice		Balance $	
2	Roll	Dice 1	Dice 2	Dice 1	Dice 2	Jhon	Peter	Jhon	Peter
3	Start	-	-	-	-	-	-	100	100
4	1	1	6	1	1	7	2	105	95
5	2	2	6	2	4	8	6	110	90
6	3	3	1	2	1	4	3	115	85
7	4	4	2	6	2	6	8	110	90

RANDBETWEEN(1; 6)   =B4+C4   =IF(F4>G4; H3+5; IF(F4=G4; H3; H3-5))

=IF(G4>F4; I3+5; IF(G4=F4; I3; I3-5))

FIGURE 3.4 Bets on the value of the sum when throwing two dice.

Example 3.2: Betting with Two Dice
John and Peter have $100 each and want to bet on dice but they do not have dice. As a substitute they decide to simulate the throwing of two dice, using six small identical chips numbered from 1 to 6. At each opportunity they bet $5 and the winner is the one who obtains the highest sum by simultaneously throwing two dice. Simulate five throws and show the final amount accumulated by each player.

Solution
Let **n** be the nth throw of both dice. $S_J(n)$ and $S_P(n)$ are the corresponding balances of John and Peter in the throw **n**. Since at the beginning both gamblers have $100, then for **n** = 0, $S_J(0) = 100$, $S_P(0) = 100$, and

$$S_J(n) = S_J(n - 1) + 5, \text{ if John wins}$$

$$S_J(n) = S_J(n - 1) - 5, \text{ if John loses}$$

- Set the number of throws to be made n = 5.
- Starting with **n** = 1, repeat until **n** = 5.
- For each throw, calculate the balance for each player.

Figure 3.4 shows the simulation of four throws obtained with the Excel sheet, after simulating the betting process.

3.10 COIN AND DICE TOSS IN EXCEL

To simulate both random events, the Excel function =RAND() returns a uniform distributed value greater or equal to 0 and less than 1, with 15 decimal digits.

$$0 \leq \text{RAND}() \leq 0.999999999999999$$

	A	B	C	D	E	F
1	Roll	Rand Coin	Coin	Dice		
2	1	0.311	Head	1		=RANDBETWEEN(1; 6)
3	2	0.614	Tail	6		
4	3	0.102	Head	2		
5	4	0.919	Tail	2		
6						
7		=RAND()	IF(B2<0.5;"Head"; "Tail")			
8						

FIGURE 3.5 Tossing a coin and a die.

All values between 0 and 1 have the same probability $(1/10^{15})$ of appearing in the cell where the function =RAND() is declared. If in addition, instead of pressing the Enter key, the **F9** key is pressed at the end of inserting the =RAND() function, the number generated becomes permanent without changing when the sheet is recalculated.

Figure 3.5 shows that cell B2 contains =RAND(), and C2 contains the function:

=IF(B2 < 0.5; "Head"; "Tail").

This relation generates a uniform random value, and then evaluates if it is less than 0.5. If B2 is less than 0.5, it writes **Head**. Otherwise, it writes **Tail**. Cells **D2:D5** contains the function =RANDBETWEEN(1; 6). This function is used to imitate the throwing of dice. It directly generates integer values between a minimum value 1 and a maximum value 6, all with equal chance in each roll.

3.11 THREE COIN TOSSES

It is desired to simulate in Excel the simultaneous tossing of three balanced coins. To run this simulation, 10, 100, 1,000, 10,000, 50,000, and 100,000 tosses will be performed. Estimate the probability of obtaining two heads, and compare the simulated estimate with the expected value. Figure 3.6 shows the formulas contained in the cells of the Excel sheet, arranged to mimic the simultaneous tossing of three coins.

	A	B	C	D	E	F	G
1	Roll	Random 1	Coin 1	Random 2	Coin 2	Random 3	Coin 3
2	1	=RAND()	=IF(B2<0.5;"Head";"Tail")	=RAND()	=IF(D2<0.5;"Head";"Tail")	=RAND()	=IF(F2<0.5;"Head";"Tail")
3	2	=RAND()	=IF(B3<0.5;"Head";"Tail")	=RAND()	=IF(D3<0.5;"Head";"Tail")	=RAND()	=IF(F3<0.5;"Head";"Tail")
4	3	=RAND()	=IF(B4<0.5;"Head";"Tail")	=RAND()	=IF(D4<0.5;"Head";"Tail")	=RAND()	=IF(F4<0.5;"Head";"Tail")

FIGURE 3.6 Formulas for the tossing of three coins.

◢	A	B	C	D	E	F	G
1	Roll	Random 1	Coin 1	Random 2	Coin 2	Random 3	Coin 3
2	1	0.058	Head	0.247	Head	0.55	Tail
3	2	0.906	Tail	0.355	Head	1.00	Tail
4	3	0.454	Head	0.027	Head	0.28	Head

FIGURE 3.7   Simulation of three coin tosses.

In Figure 3.6 in cell B2, the Excel function =RAND() generates a uniform random value which imitates the result obtained when the first coin is tossed. The formula contained in cell **C2** checks if the value enclosed in the previous cell, **B2**, is less than 0.5 to display the text "Head". If not, the text "Tail" will be displayed in cell C2. Cells **D2**, **E2**, **F2**, and **G2** imitate the result of tossing the other two coins. All are for the first roll.

Figure 3.7 shows the result obtained by the formulas shown in Figure 3.6 after simulating three coin tosses.

Figure 3.8 contains the function =**COUNTIF**(). This function displays the formulas included in the Excel cells to count the number of **Head** texts contained in the row that imitates the first pitch. That is, the cells from C2 to G2. Then the function =**IF**, determines whether the count of **Head** in cell **H2** (Figure 3.8) is equal to 2, so it will store a value of 1 in cell **I2**. Although if the **Head** count in cell H2 is not 2, a zero will be placed in cell I2.

The formula =**SUM**(I2; I10001), arranged in cell K2, sums all the 1s, or favorable events, corresponding to the achievement of exactly two Heads contained in the range of cells **I2** to **I10001**. That is, it counts the number of favorable events (two Heads) in a simulation of ten thousand throws. Finally, using the formula =**K2/10000**, the probability of obtaining two Heads is estimated by dividing the number of favorable events or the number of times that the sum of Heads was 2, by the 10,000 throws made.

Figure 3.8 shows in cell L2 the result of simulating only three throws. In this case, one favorable event is obtained in three attempts from which the probability of obtaining exactly two Heads is estimated as:

Probability(2 Heads)) = Favorable Cases / Possible Cases = 1/3

◢	A	B	C	D	E	F	G	H	I	J	K	L
1	Roll	Random 1	Coin 1	Random 2	Coin 2	Random 3	Coin 3	Heads / Roll	Place 1 If 2 Heads		Total Favorable Events	Probability of getting 2 Heads
2	1	0.980	Tail	0.095	Head	0.96	Tail	1	0		1	0.333
3	2	0.642	Tail	0.172	Head	0.52	Tail	1	0			
4	3	0.368	Head	0.899	Tail	0.03	Head	2	1		=SUM(I2:I4)	=K2/3
5												
6						=COUNTIF(C2:G2; "Head")		=IF(H2=2; 1; 0)				

FIGURE 3.8   Estimating the probability of getting two Heads.

TABLE 3.6 Estimated probability for **n** shots.

SAMPLE SIZE N (NUMBER OF THROWS)	SIMULATED (ESTIMATED) PROBABILITY $\hat{p}$
10	0.500
100	0.420
1,000	0.396
10,000	0.377
50,000	0.37292
100,000	0.37454

Since this is a random sampling procedure, the result obtained varies from one sample to another. For three tosses, the estimate may give 0.66 at one time, 0.33 at another, 0 at another, or even 1. Achieving an acceptable approximation for the probability of obtaining two heads when tossing three coins simultaneously requires simulating approximately 10,000 tosses. Table 3.6 shows the estimate of the referred probability for different number of tosses. There is a tendency to estimate with greater accuracy as the number of tosses increases, i.e., as the sample size to be simulated increases.

3.12 PROBABILITY OF OBTAINING TWO HEADS WHEN TOSSING THREE COINS

The probability of obtaining two heads when tossing three coins corresponds to a binomial process with parameters: probability of success p = 0.5; number of attempts n = 3; and value of the variable to be calculated x = 2.

$$P(2 \text{ heads}) = P(x = 2) = \frac{n!}{x!(n-x)!}p^x(1-p)^{n-x}$$

$$P(x = 2) = \frac{3!}{2!(3-2)!}0.5^2(1-0.5)^{3-2} = 0.375$$

Binomial probability can be calculated using the following Excel statistical function:

=**BINOM.DIST**(successes x; attempts n; probability of success p; 1 if cumulative).

In this case, the probability of obtaining exactly two faces in three attempts, without accumulation of probabilities, is achieved by:

=**BINOM.DIST**(2; 3; 0.5; 0) = 0.375

When comparing the simulated values obtained with the exact value of the probability, it is observed that the best approximation is achieved as the number of throws increases (Table 3.6).

The accuracy and reliability of the estimate depend on the desire of the researcher, defined according to the sample size.

To estimate the parameter of the probability of getting exactly two heads when tossing three coins, with an accuracy of 1,000th of an error and a confidence level of 95%, it is required to execute at least 240,100 tosses as determined below:

If the estimation is conditioned on an error with 1,000th of a unit (0.001), at 95% confidence, one obtains:

$$n \geq \frac{z_{0.95}^2}{4 * \epsilon^2} = \frac{1.96^2}{4 * 0.001^2} = 240,100 \text{ tosses}$$

The sample size relationship above allows us to estimate the number of throws required to obtain a probability estimate close to the theoretical binomial value. Its components are described later in Section 5.2 of Chapter 5.

In summary, a simulation experiment where 240,100 tosses of three coins are performed allows us to estimate the probability of obtaining two heads, so that the interval [0.374; 0.376] in 95 out of 100 times will contain the expected probability value. That is, our simulated sample statistic will be within 1,000th of error for the exact analytical value of the population parameter during 95% of the time. It is 0.375. Assuming an estimation error of 100th, i.e., $\varepsilon = 0.01$, we can determine the sample size **n** and conclude that it is reduced to only 2,401 throws.

$$n \geq \frac{z_{0.95}^2}{4 * \epsilon^2} = \frac{1.96^2}{4 * 0.01^2} = 2,401 \text{ tosses}$$

Generation of Random Variables

4

4.1 GENERATION OF RANDOM VARIATES

A simulation model mimics the behavior of an existing or proposed system. Systems are observed in their existential states, defined by their entities interactions and events. Various events could occur in the system over time. Some of these events are random in nature and must be represented by appropriate probability models.

For example, in the emergency department of any hospital, the number of patients arriving in a day is random. The time between the arrival of one patient and the next has unplanned behavior called stochastic or random inter-arrival time. The number of patients at the start and the inter-arrival time records can be represented by various probability distributions such as Poisson, Exponential, Gamma, Lognormal, Weibull, Beta, etc. Once a patient arrives, his or her permanence in the system must be evaluated. Some patients require hospitalization, while others once treated will return to their origin. A Bernoulli distribution can model the historical percentage of patients requiring hospitalization.

In summary, simulation requires procedures to mimic the occurrence of uncertain events symbolized by random variables represented in probabilistic models. Probability models are the quantitative tools available to the observer to measure the likelihood of an unpredictable event occurring.

4.2 GENERATION OF BERNOULLI RANDOM VARIABLES

A random process that shows only one of two possible outcomes can be modeled by a Bernoulli random variable. Let **x** be the Bernoulli random variable whose two likely values are Success or Failure (Not Success). If the probability of success is **p** and the probability of failure is **1-p**, then the procedure for simulating process outcomes is shown in Table 4.1.

DOI: 10.1201/9781032701554-4

TABLE 4.1   Procedure for simulating a Bernoulli event

1. Let **p** be the probability of event success. Generate a uniform random value **R**
2. If **R** < **p**, then the variable **x** = "Success"
3. Otherwise, **x** = "Not Success"

Example 4.1: Quality Control

A quality control inspector samples a manufacturing line of a sophisticated electronic product. It is known that the proportion of defective equipment is approximately 12%. If the inspector samples three pieces of equipment, simulate the condition of each of the sampled pieces of equipment.

Solution

Let **p** = 0.12 be the probability that an inspected piece of equipment is defective (Success). To simulate the process of inspecting three pieces of equipment, three uniform random values are generated, one for each piece of equipment. The random **R** values are values between 0 and 1. Since all **R** values have the same probability of occurrence, 12% of the values are assigned to the random event Success. That is, each random value **R**, which is less than 0.12 corresponds to the selection of a defective equipment. If the **R** number is greater than or equal to 0.12, then the product is not defective and is Passed (Table 4.2).

TABLE 4.2   Quality control

EQUIPMENT NO.	R RANDOM	EQUIPMENT CONDITION (VALUE OF VARIABLE X)
1	0.482	Passed
2	0.101	Defective
3	0.756	Passed

4.3 GENERATION OF BINOMIAL RANDOM VARIABLES

A binomial random variable of parameters **p** and **n** corresponds to the repetition of **n** independent Bernoulli processes, where the fixed probability of success in each repetition is **p**. The value of the random variable **x** corresponds to the number of favorable events obtained by repeating **n** events. Its probability function is:

$$P(x) = \frac{n!}{x! * (n-x)!} * p^x * (1-p)^{n-x}, \quad x = 0, 1, 2, \ldots, n$$

The procedure consists of generating **n** uniform random values **R**, and accumulating in the variable **x** the number of times that the random value **R** is less than **p**. That is, repeat the Bernoulli process **n** times and count in **x** the number of successes. By the way, when **n** tends to infinity and the probability of success **p** tends to zero, the Binomial distribution tends to an average Poisson $\lambda = n*p$.

Example 4.2: A Packaging Operation
In a packaging process, a random sample of three jars content is measured every 30 minutes and the number of jars weighing less than 900 grams is counted. If a canister weighs less than 900 grams it is considered a Fail (F). Otherwise, it is Accepted (A). It is known that 20% of the jars weigh less than 900 grams. Simulate 2 hours of sampling of the packaging process.

Solution
The time between sampling corresponds to the selection of each sample, so that each choice corresponds to a time of 30 minutes. Let **p** = 0.20 be the probability that a container encloses less than 900 grams. Table 4.3 presents the result of the simulated sampling sequence. The three R_i columns record the random value between 0 and 1. If the value of R_i is less than 0.20, then it means the container contains less than 900 grams and is recorded with **F** (Failure). If R_i is greater than or equal to 0.20 then the canister complies and its weight is accepted. It is recorded with **A** (Acceptance).

TABLE 4.3 Binomial generation. Packaging process

N	R_1	CONTAINER 1	R_2	CONTAINER 2	R_3	CONTAINER 3	WEIGHT <900 GRAM
1	0.14	F	0.05	F	0.45	A	2
2	0.67	A	0.42	A	0.98	A	0
3	0.45	A	0.77	A	0.21	A	0
4	0.23	A	0.73	A	0.16	F	1

4.4 GENERATION OF GEOMETRIC RANDOM VARIABLES

When performing a series of independent Bernoulli tests with probability of success **p**, the geometric random variable **x** counts the number of attempts made until the first success is obtained. For example, the number of parts inspected until the first defective one is found. Its probability function is:

$$f(x) = p * (1 - p)^x$$

For x = 0, 1, 2, … of average and variance:

$$E(x) = \frac{1 - p}{p} \quad \text{and} \quad V(x) = \frac{1 - p}{p^2}$$

The generating expression for geometric random values is as follows:

$$x = \text{integer}\left[\frac{Ln(R)}{Ln(1 - p)}\right] \tag{4.1}$$

where **R** is a uniform random value between 0 and 1. Its implementation in the Excel sheet is done with the following function:

=ROUNDDOWN(LN(RAND())/LN(1-p); 0).

The function **ROUNDDOWN** rounds the result of the quotient indicated in the expression (4.1) to zero decimal places which guarantees for answer an integer value.

Example 4.3: Quality Inspection
A quality inspector checks production batches on a daily basis. The probability of finding a defective part is 0.07. Each day the officer inspects until he finds a defective part. Simulate for 4 days and determine the number of products the inspector must check daily until the first defective part of the day is removed.

Solution
We have p = 0.07. In Figure 4.1, the cells of the range B2:B5 contain the function

=ROUNDDOWN(LN(RAND())/LN(1–0.07); 0)

which allows us to simulate the number of daily parts checked until the first defective part is found. This function requires two arguments: (a) the expression to round and (b) the number of decimal places to display in the result. In this case, the function takes the integer part of the quotient of both logarithmic expressions.

	A	B	C
		Products check until	
1	Day	first defective	
2	1	5	
3	2	11	
4	3	7	
5	4	20	
6			
7		=ROUNDDOWN(LN(RAND())/LN(1-0.07);0)	

FIGURE 4.1   Geometric generation: part inspection.

4.5 GENERATION OF POISSON RANDOM VARIABLES

The random variable **x** of the Poisson distribution symbolizes the number of events occurring in a time span, area, or volume. Its mean or average is usually denoted by λ and represents the **rate value of occurrences of the event per unit of time, area, or volume**.

For values λ close to or greater than 20, the shape of the distribution becomes more symmetrical, tending to a normal distribution. Its variable **x** is a discrete count variable which only takes integer values. For example, **cars** serviced **per hour**; number of **accidents** occurring on a highway **per day**; number of **defects** in a 50-meter **roll** of cloth; number of **calls** received **per hour** at a telephone exchange; number of radioactive **decays/year**, number of **gas leaks** in a **kilometer** of pipeline, etc.

All these cases can be described by a discrete random variable which takes non-negative integer values. This distribution is very frequent in Operations Research applications, especially in the area of Queuing Theory.

The probability density function of the Poisson variable **X** is:

$$P(X) = \frac{\lambda^X e^{-\lambda}}{X!}, \ X = 0, 1, 2, \ldots \text{ and } \lambda > 0$$

The Poisson variable is characterized by having **equal mean and variance** with value λ. That is to say: $E(x) = V(x) = \lambda$.

With this distribution we estimate the probability that the event of interest on average λ occurs exactly **X** times during a certain period of time. **X** can also represent the number of occurrences over an area or volume; for example, the number of imperfections in a sheet of zinc.

Its application requires that the event it expresses meets the following conditions:

1. The probability of occurrence of more than one event in an infinitesimal amount of time is 0.
2. Memoryless process: The probability of an event occurring in a period of time is independent of what has occurred previously.
3. The probability of an event occurring during a time interval is proportional to the length of the interval.

Generally, the random variable of the Poisson distribution measures the number of times an event occurs during a period of time. Its mean represents the average value of occurrences per unit of time. For example, 10 **vehicles** arrive **per hour**. This is equivalent to a random exponential average inter-arrival time of 6 minutes between the arrival of one car and the next; 6 rains/month, 30 pieces/minute. If average is Poisson where 12 babies are born/day, it is equivalent to say that it is expected one baby each exponential time with mean 2 hours. Always number of **events**/time by **Poisson** and **time**/events by **Exponential**.

4.5.1 Poisson Random Generation Methods

To generate Poisson random values with average λ, the following algorithm can be used (Table 4.4), which executes a cumulative multiplication on the variable **P**:

TABLE 4.4   Method for generating Poisson random variables

1. Start $X = 0$, and $P = 1$
2. Calculate $Z = e^{-\lambda}$
3. Generate a random uniform R(0,1)
4. Calculate $P = P * R$
5. If $P <= Z$, terminate and go to step 7
6. Increment $X = X + 1$ and return to step 3
7. The random Poisson value is X.

Recall that the neutral value of the arithmetic operation sum is 0. In the case of multiplication, the neutral element is 1. Since the algorithm accumulates summation in **X** and the product operator for n variables in **P**, it starts accumulating with $X = 0$ and $P = 1$.

Example 4.4: Parts Manufacturing
The number of defective parts per hour in a production line is a Poisson variable of average 2. Simulate 2 hours of production to determine the number of defective parts produced in each hour.

Solution
The mean value $\lambda = 2$ allows to evaluate $z = e^{-2} = 0.135$. Initialize $X = 0$ and $P = 1$. The first random value $R = 0.38$ is generated. The result of its product P is 0.38, which is greater than 0.135; en consequence, we increase X by 1, and we proceed to generate another $R = 0.5$. The product of $P = 1*0.38*0.5$ equals 0.19 and this product still greater than 0.135, which increases X to 2 and requires generating another R. Since the new product accumulated in $P = 0.1178$ is less than $Z = 0.135$, the count is terminated, leaving $X = 2$ defective parts for first hour.

Table 4.5 records the sequence of events. In the first hour, two defective parts were obtained. In the second hour zero parts; the first random number R for the second hour was 0.11. It is less than 0.135, a value of e^{-2}, which terminates the algorithm and leaves the value of X at zero defects during the second hour.

TABLE 4.5 Defective parts per hour

HOUR	X DEFECTIVES	P	λ	$Z = e^{-2}$	RANDOM UNIFORM	PRODUCT
1	0	1	2	0.135	0.38	1*0.38
	1	0.38			0.50	1*0.38*0.5
	2	0.19			0.62	1*0.38*0.5*0.62
		0.1178*				
2	0	1	2	0.135	0.11	1*0.11
		0.11*				

Note
* Value less than 0.135.

4.5.2 Poisson Generation with Excel Functions

To generate Poisson random values of average λ using directly the Excel sheet, two options are available:

=**BINOM.INV**(λ / 0.001; 0.001; RAND())

Another alternative is from the toolbar by clicking:

DATA / Data Analysis / Random Number Generation

Fill in the format of the tab indicating the following: (a) **Number of Variables**: number of columns you want to fill with Poisson values; (b) **Number of Random Numbers** required for each column; (c) select in the list **Poisson** distribution; (d) **Lambda**: average Poisson lambda; (e) **Output Range**: indicate an output range for the results. Finish with OK.

If the **Data Analysis** option does not appear in the **Data** toolbar, it is necessary to activate it using the following sequence:

File / Options / Add-Ins / Analysis Tool-Pak / Go / Analysis Tool-Pak / OK.

Example 4.5: Customer Claims

Use Excel to generate five Poisson values corresponding to the daily claims for poor service received by an insurance company. The number of claims is a Poisson variable averaging 3.5 claims/day.

Solution
Figure 4.2 shows the use of the function =BINOM.INV() to simulate Poisson values.

	A	B	C
1	**Day**	**Claims/Day**	
2	1	2	
3	2	1	
4	3	3	
5	4	3	
6	5	4	
7			
8	=BINOM.INV(3.5/0.001; 0.001; RAND())		

FIGURE 4.2 Poisson generation: claims/day.

The function =BINOM.INV(3.5/0.001; 0.001; RAND()), enclosed in the range of cells B2:B6, generates the random values representing the number of daily claims received by the insurer averaging 3.5.

To generate five Poisson values using the **Data Analysis** option, the format is filled in as shown in Figure 4.3.

Random Number Generation ? ✕

Number of Variables:	1	OK
Number of Random Numbers:	5	Cancel
Distribution:	Poisson ⌄	Help

Parameters

Lambda = 3.5

Random Seed:

Output options
- ⦿ Output Range: B2.B6
- ○ New Worksheet Ply:
- ○ New Workbook

FIGURE 4.3 Data analysis. Poisson generation of mean 3.5.

4.6 GENERATION OF HYPERGEOMETRIC VARIABLES

The Hypergeometric variable counts the number of objects **x** with a characteristic of interest (success), obtained by extracting without replacement a sample of **m** objects, from a population containing **N** objects, among which there are **E** objects with the characteristic of interest (Success). That is to say, **x** counts the number of successes obtained in the extracted sample of size **m**. The sample space or possible values of **x** is the set of the integers from 0 to **m**, or from 0 to **E**, if **E** < **m**. In summary, a variable has a Hypergeometric distribution if it comes from an experiment that meets the following conditions:

1. A sample of size **m** is taken without replacement from a finite set of **N** objects.
2. The **E** objects of the total of **N** possess a characteristic of interest, which allows them to be classified as **E** successes, and the remaining, **N** – **E** objects as failures. In this case, the probability of Success in successive extractions is not constant, because it depends on the result of the previous exits. Therefore, the draws are dependent. The input parameters to the distribution are **m**, **E**, and **N**.

The probability function of the Hypergeometric variable is as follows:

$$H(x, \ m, \ E, \ N) = \frac{\left[\begin{matrix} E \\ x \end{matrix}\right]\left[\begin{matrix} N - E \\ m - x \end{matrix}\right]}{\left[\begin{matrix} N \\ m \end{matrix}\right]}, \ x = 0, \ 1, \ 2, \ ...,m$$

The values of the mean and variance are calculated according to:

$$\text{Mean} = E(x) = \frac{m * E}{N}$$

$$\text{Variance} = V(x) = \frac{m * E * (N - m)}{N * (N - 1)}\left(1 - \frac{E}{N}\right)$$

4.6.1 Hypergeometric Generator in Visual Basic

Below is a program (Macro) in Visual Basic, which allows to generate random values for the parameters of any Hypergeometric distribution. The macro is ready to be used inside an Excel sheet. Table 4.6 shows the procedure to follow to take it to Excel. The macro is shown in Table 4.7.

TABLE 4.6 Procedure to create Hypergeometric macro

If the **DEVELOPER** tab does not appear in the ribbon, it is necessary to activate it. To do so:
Click **FILE / Options / Add-Ins / Analysis ToolPak – VBA / Go** ... **/ Analysis ToolPak – VBA / OK**
1. Select **DEVELOPER**
2. In the Options Ribbon select **Macros Macro name**: Hypergeometric
3. **Click** Create
4. Type all the instructions for the Hypergeometric Sub indicated in Table 4.7

(Continued)

TABLE 4.6 *(Continued)* Procedure to create Hypergeometric macro

5. Select **Execute**. The result of the execution can be seen in the Excel sheet. If you look at the bar or ribbon at the bottom of the computer screen, Microsoft Excel and Microsoft Visual tabs appear. By clicking on one or the other, you can exit and enter to run or view results.
6. To save the Excel sheet and its macro click the Office button. Then, Save as Excel Macro Enabled workbook.

TABLE 4.7   Procedure VB for sub Hypergeometrics

```
Sub Hypergeometric()
Range("a2.c1000").ClearContents
Cells(1, "A") = "Population Size (N)"
Cells(1, "B") = "Number of Objects of Interest in the Population (E)"
Cells(1, "C") = "Number of Objects to Extract (m)"
N = Val(InputBox("Size of the Population to Sample?N = ", "Size?"))
Cells(2, "A") = N
E = Val(InputBox("Objects in the Population? E =", "Objects of Interest"))
Cells(2, "B") = E
m = Val(InputBox("How many Objects to Extract? m = ", "Sample Size"))
Cells(2, "C") = m
Cells(4, "A") = "Sampling No."
Cells(4, "B") = "Number of Objects of Interest Extracted (x)"
N1 = N
E1 = E
K = Val(InputBox("How many Values to generate?", "Number of values?"))
For j = 1 To K
  x = 0
  N = N1
  E = E1
  For i = 1 To m
    Prob = E / N
    R = Rnd()
    If R < Prob Then
      x = x + 1
      E = E - 1
    End If
    N = N - 1
  Next i
  Cells(j + 4, "A") = j
  Cells(j + 4, "B") = x
Next j
End Sub
```

When copying make sure that only one Hypergeometric Sub reference appears and only one End Sub closes the macro.

Example 4.6: Travelers and False Passports

A customs officer has been informed about the number of travelers with false passports arriving on a certain flight that always carries 80 passengers. With certainty about 15%

of the passengers have false documentation. Due to time constraints, the officer decides to check 10 passengers. Simulate 5 times and obtain the corresponding value of the Hypergeometric variable, which reflects the detected number of passengers with false passports for each of the five opportunities.

Solution
With $N = 80$, $E = N * 0.15 = 12$ and $m = 10$, obtain five values from a Hypergeometric distribution that personifies the passengers with false passports.

Figure 4.4 shows the input requirements for the parameters of the distribution when running the Hypergeometric Excel Macro.

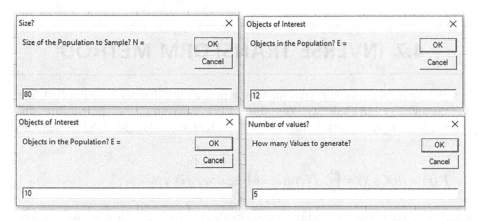

FIGURE 4.4 Data entry when running macro Hipergeometric.

Figure 4.5 in its column B gives the values of the Hypergeometric random variable corresponding to the input parameters. This macro can generate thousands of Hypergeometric random values.

	A	B	C
	Population Size (N)	Number of Objects of Interest in the Population (E)	Number of Objects to Extract (m)
1			
2	80	12	10
3			
4	Sampling No.	Number of Objects of Interest Extracted (x)	Click to generate K Hipergeometrics
5	1	2	
6	2	0	
7	3	2	
8	4	1	
9	5	4	

FIGURE 4.5 Hypergeometric: Passengers with false passport.

Figure 4.5 shows the use of a button arranged to execute the Hypergeometric value generating macro. A button is an element belonging to the form controls. To define it, in the Excel sheet, select the option **Programmer / Insert** from the toolbar and select Button from the Form Controls tab. You can see that the cursor changes to a crosshair, which must be positioned on the spreadsheet. When left clicking the Assign Macro tab appears. For macro name, **Select** Hypergeometric / **OK**. Subsequently, to assign a name or caption to the control button, **right click** it on the mouse and **Modify** text. When the button is clicked its associated macro is executed.

4.7 INVERSE TRANSFORM METHOD

If **x** is a random variable represented by a probability density function **f(x)**, then its **cumulative probability function F(x)** measures the probability that the variable **x** takes any value below a reference value **X**. **F(x)** is obtained according to:

$$F(x) = P(X \le x) = \int_{-\infty}^{x} f(x)dx, \quad \text{where } 0 \le F(x) \le 1$$

Since the cumulative probability values, like the values generated by the RAND() function, are values between 0 and 1; one can generate a RAND() value **R**, make it correspond to a probability value and determine the value of the variable **x** associated with that probability value. It is, equate the RAND() **R** to the cumulative function **F(x)** and then clear **x** as a function of **R**.

Performing the clearing of **x** is called finding the inverse transform. The general inverse transform procedure is shown in Table 4.8.

TABLE 4.8 General inverse transform procedure

1. With f(x) being the target random expression, from the probability density function **f(x)**, obtain the cumulative probability function **F(x)**.
2. Generate a random value **R** and make **R = F(x)**
3. To clear the variable **x** as a function of **R**, i.e., to find the inverse transform: $x = F^{-1}(R)$
4. The new function obtained can be used to generate random values with probability density function *f(x)*.

Next, the inverse transform method is applied to find generating procedures for uniform and exponential random variables.

4.8 GENERATION OF UNIFORM RANDOM VARIABLES

The uniform probability distribution function defined by its minimum **a** and maximum **b** parameters has the following probability density function:

$$f(x) = \frac{1}{b-a} \text{ only for } a \le x \le b$$

With mean value E(x) = (a + b)/2 and variance V(x) = (b – a)^2/12.

By applying the inverse transform method to the uniform probability function according to the procedure expressed in Table 4.5, we obtain the following:

$$F(x) = \int_a^x \frac{1}{b-a} dx = \frac{x-a}{b-a}$$

Equating:

$$F(x) = R = \frac{x-a}{b-a}$$

where **R** is a uniform random value between 0 and 1.

The uniform generator is achieved after clearing x. In summary, to generate uniform random values of minimum parameter **a,** and maximum parameter **b,** a uniform random **R** value of parameters 0 and 1 is obtained and the value of the uniform random variable is calculated by means of:

$$x = a + (b-a) * R$$

Any generated **x** value will be between the limiting values **a** and **b** and it is distributed according to a uniform probability density function.

Example 4.7: Operator Times
The time taken by an operator to perform a task corresponds to a uniform random variable with parameters between 10 and 18 minutes. Simulate the execution of three tasks performed by the operator (Table 4.9).

TABLE 4.9 Manual generation of three uniform times

TASK NO.	R	DURATION (MIN.) X = 10 + 8*R
1	0.245	11.96
2	0.872	16.98
3	0.487	13.90

Solution

With **a** = 10 and **b** = 18 the time **x** is obtained by:

$$x = 10 + (18 - 10) * R$$

where R is a uniform random number between [0; 1).

4.8.1 Generation of Uniform Variables in Excel

To obtain uniform random values between **a** and **b** from the Ribbon or Toolbar, Excel offers the option:

Data / Data analysis / Random number generation / Distribution / Uniform.

Additionally, the function **=RANDBETWEEN** (minimum value; maximum value) generates uniform integer random values between the minimum and maximum values. For example, the expression **RANDBETWEEN**(−50; 100) repeated 5 times, generates five random values in the indicated range, each with equal probability of occurrence although strictly integers.

In Figure 4.6, the range of cells **B2:B6** contains the expression =RANDBETWEEN (−50; 100).

	A	B
1	No.	Uniform Random (-50,100)
2	1	98
3	2	-27
4	3	79
5		
6		=RANDBETWEEN(-50;100)
7		

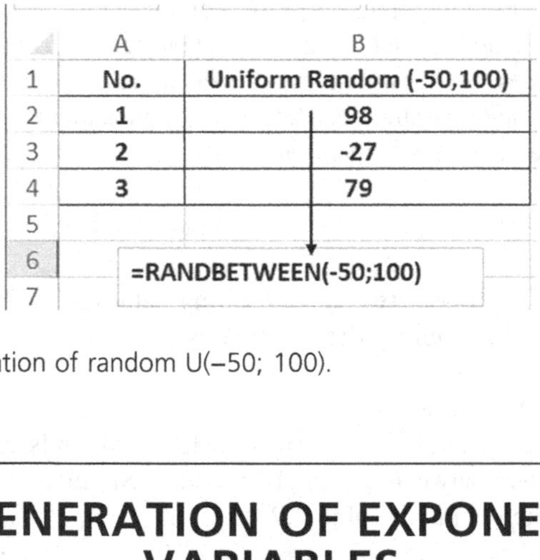

FIGURE 4.6   Generation of random U(−50; 100).

4.9 GENERATION OF EXPONENTIAL VARIABLES

The exponential probability distribution can be used to represent the distribution of the time of occurrence of two independent events with average $1/\beta$. For example, the time that elapses between the consecutive arrival of two cars at a toll booth, the duration time of a light bulb, or the time between the arrival of ships at a dock.

The exponential probability distribution function **f(x)** defined by its mean $\frac{1}{\beta}$ has the following model:

$$f(x) = \frac{1}{\beta}e^{\frac{x}{\beta}} \text{ For } x \geq 0 \text{ and } \beta > 0$$

On average: $E(x) = \frac{1}{\beta}$ and $V(x) = \frac{1}{\beta^2}$, where β corresponds to the average number of occurrences of the event per unit of time, area, or volume and $\frac{1}{\beta}$ the average time between occurrences of the event. By applying the inverse transform method, we obtain:

$$F(x) = \int_0^x \frac{1}{\beta}e^{x/\beta}dx = 1 - e^{-\beta x}$$

$$R = 1 - e^{-\beta x}$$

where R is a uniform random value between 0 and 1:

$$x = -\frac{1}{\beta}\ln(1 - R)$$

Since R and $(1 - R)$ are complementary uniform random values, the factor $1 - R$ can be substituted for R. The value of **x** corresponds to the value of the exponential variable of average $\frac{1}{\beta}$. In consequence, the generator for a random exponential variables with average $\frac{1}{\beta}$ is given by:

$$x = -\frac{1}{\beta}\ln(R)$$

Example 4.8: Lifetime of an Electronic Device

The lifetime of an electronic device is random exponential averaging 1,000 hours. Five devices are sold. If a device lasts less than 200 hours, the buyer is reimbursed with an equivalent device. Simulate the lifetime of the five electronic devices and determine the number of devices replaced.

Solution

The average device lifetime is: $\frac{1}{\beta} = 1,000$ hours

According to the result shown in Table 4.10, one of the five equipment purchased needs to be replaced.

TABLE 4.10   Exponential random lifetimes

DEVICE NO.	RANDOM R	LIFETIME (HOURS) $-1,000*LN(R)$	REPLACE?
1	0.2059	1,580	No
2	0.6747	393	No
3	0.2096	1,562	No
4	0.8465	167	Yes
5	0.1116	2,192	No

4.10 RELATIONSHIP BETWEEN EXPONENTIAL AND POISSON DISTRIBUTIONS

The Poisson and Exponential distributions are allied and share their average in an inversely relationship. It means, where there is one exponential with mean $\frac{1}{\beta}$ then there is too a Poisson with mean β and vice versa. For example, in a supermarket, if the number of customers per unit of time arriving at a checkout counter is assumed to be Poisson with average $\beta = 30$ customers/hour, then the time between the arrival of one customer and the next is exponential with average $\frac{1}{\beta} = 0.0333$ hours/customer = 2 minutes/customer.

The Poisson variable is a counter of discrete values: 0, 1, 2, ... Its average measures:

$$\frac{number\ of\ events\ occurred}{time}$$

While the exponential variable is continuous, and its average measures:

$$\frac{time}{number\ of\ events\ occurred}$$

That is for example, if in a region the amount of rainfall per month is modeled as a Poisson variable of average $\beta = 6$ rainfalls/month; it means 6 rainfalls/30 days; it means that it is expected to have one rainfall every 5 days. Then, in that region, the average time between rains corresponds to an exponential variable of average $\frac{1}{\beta} = 5\frac{days}{rainfall}$. In summary, in the Poisson-Exponential marriage, the average of one variable corresponds to the inverse of the average of the other.

4.11 STANDARD NORMAL RANDOM GENERATION

Standard normal random values are used to generate normally distributed random values agreeing to the desired mean and variance. According to the Central Limit Theorem, the sum of **n** random variables R_i yields a normal random variable of mean **n∗μ** and variance $n * \sigma^2$. In the following relation:

$$x = \sum_{i=1}^{i=n} R_i$$

where R_i corresponds to a uniform random value $[0,1)$, of known mean 0.5 and variance 1/12. Then, by standardizing the R_i values we obtain:

$$x = \frac{\sum_{i=1}^{i=n} R_i - \mu}{\sigma} = \frac{\sum_{i=1}^{i=n} R_i - n * 0.5}{\sqrt{\frac{n}{12}}}$$

Taking for convenience and simplification **n** = 12 uniform R values, the following relation is achieved. It is valid for generating standard normal random values of mean 0 and variance 1:

$$R_{(0,1)} = \sum_{i=1}^{i=12} R_i - 6$$

Example 4.9: Standard Normal Random Generation

Generate three standard normal random values.

	A	B	C	D	E	F	G	H	I	J	K	L	M	N
1	Value	R1	R2	R3	R4	R5	R6	R7	R8	R9	R10	R11	R12	Standard Normal Random R[0,1)
2	1	0.2672	0.0776	0.5326	0.2608	0.5759	0.2991	0.1356	0.0539	0.9503	0.2114	0.0610	0.1133	-2.4612
3	2	0.5895	0.3581	0.9046	0.2989	0.5443	0.8759	0.9915	0.0237	0.9313	0.9463	0.4141	0.1531	1.0313
4	3	0.7487	0.2609	0.7864	0.1957	0.6547	0.6237	0.3760	0.1957	0.9940	0.5192	0.8636	0.4130	0.6315
5														
6		=RAND()												=SUM(B2:M2) - 6

FIGURE 4.7 Standard normal random generation.

After applying the generating relation R(0,1) from 12 uniform random values between [0, 1), R_1, R_2, ..., R_{12}, adding them and subtracting 6, the first normal random value is obtained: $R_1(0,1) = -2.4612$ (Figure 4.7). By repeating the relationship, the second value $R_2(0.1) = 1.0313$ is obtained. A third value 0.6315, and so on. Another alternative option for creating standard normal random numbers is to use the Box-Muller algorithm which is detailed below.

4.11.1 Box-Muller Procedure

This method obtains two values N(0,1) from two uniform random values R[0,1) according to the steps given in Table 4.11.

TABLE 4.11   Box-Muller procedure.

Obtain two uniform random $R_{1(0,1)}$ and $R_{2(0,1)}$
Calculate:

$$X_1 = \sqrt{-2 * log(R_{1(0,1)})} * cos(2 * \pi * R_{2(0,1)})$$

$$X_2 = \sqrt{-2 * log(R_{1(0,1)})} * sin(2 * \pi * R_{2(0,1)})$$

The values X_1 and X_2 are normal random values of mean = 0 and variance = 1. Log corresponds to the natural or neperian logarithm with base e.

Example 4.10: Box-Muller Random Generation

Apply Box-Muller to obtain six standard normal random values (Figure 4.8).

Solution

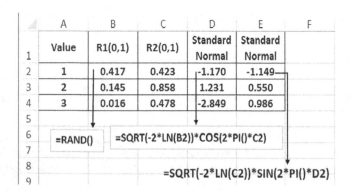

FIGURE 4.8   Standard normals with Box-Muller.

4.11.2 Random Normals with Polar Method

This procedure is an alternative to generate standard normal random values (Table 4.12).

TABLE 4.12   Polar method for standard normal random.

1. Obtain two uniform random R_1 and R_2.
2. Calculate: $V_1 = 2*R_1 - 1$ and $V_2 = 2*R_2 - 1$
3. Calculate: $Z = V_1^2 + V_2^2$
    If $Z < 1$ then

$$X_1 = \sqrt{\frac{-2*\log(Z)}{Z}} * V_1$$

$$X_2 = \sqrt{\frac{-2*\log(Z)}{Z}} * V_2$$

Otherwise

    If $Z \geq 1$ return to step **1**.

Example 4.11: Polar Method

Apply the polar method to obtain at least two standard normal random values (Figure 4.9).

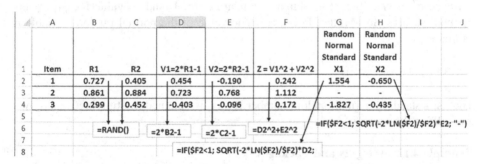

FIGURE 4.9   Standard normals. Polar method.

4.11.3 Standard Normals with Excel

The Excel statistics function **=NORM.S.INV**(probability) directly generates standard normal random values of **mean 0** and **variance 1**. The probability argument can be the uniform random function =RAND() (Figure 4.10).

	A	B
1	Item	Random Normal Standard N(0,1)
2	1	1.52808
3	2	0.16734
4	3	0.40186
5	4	-1.96570
6		
7		=NORM.S.INV(RAND())

FIGURE 4.10   Standard normals with Excel.

Standard normal random values are used to generate standard normal random values. To do this, the standard value $N_{(0,1)}$ is multiplied by the normal deviation σ and the normal mean μ is added to it: $\mu + \sigma * N_{(0,1)}$.

4.12 GENERATION OF NORMAL RANDOM VARIABLES

The normal probability distribution of mean μ and variance σ^2 has density function:

$$f(x) = \frac{1}{\sigma\sqrt{2\pi}} e^{\frac{-(x-\mu)^2}{2\sigma^2}}, \; for \; -\infty \leq x \leq \infty$$

If **R** is a normal random variable of mean $\mu = 0$ and variance $\sigma^2 = 1$, then

$$x = \mu + \sigma * R_{N(\mu,\sigma^2)}$$

To generate random variables distributed according to a normal of mean μ and variance σ^2, it is sufficient to obtain a standard normal random value $R_{N(0,1)}$ by any method, such as Box-Muller, Polar, etc. The value of the normal random variable is estimated as:

$$x = \mu + \sigma * R_{N(o,1)}$$

That is, a standard normal random number $R_{N(0,1)}$ is obtained multiplying $R_{N(0,1)}$ by σ and adding the mean μ.

Example 4.12: Weights and Packages
The weight in grams of a package containing tuna fish is distributed according to a normal with an average of 400 grams and a standard deviation of 30 grams. Simulate and record the weight of three packages (Table 4.13).

Solution

TABLE 4.13   Generation of the weight of three containers.

CONTAINER NO.	RANDOM NORMAL $R_{N(0,1)}$	WEIGHT $= 400 + 30 * R_{N(0,1)}$ (GRAMS)
1	−0.125	396.3
2	0.806	424.2
3	−0.738	377.9

4.12.1 Random Normals with the Excel Function

Excel offers two options to generate normal random values: one from the toolbar, via the **Data / Data analysis** tab; and the other option via the inverted normal function. Its form is:

= **NORM.INV**(RAND(); μ; σ).

For example, to generate in any Excel cell a normal random value of mean μ = 75 and standard deviation σ = 15, it is arranged in the cell:

= **NORM.INV**(RAND(); 75; 15).

The expression: =75+15∗**NORM.S.INV**(RAND()) also generates normal random value of average 75 and standard deviation 15.

Example 4.13: Direct Normals in Excel

Obtain three random normal values of average 75 and deviation 15.

Solution
The range of cells **B2:B4** contains the statistical function **NORM.INV**(RAND(); 75; 15) (Figure 4.11).

	A	B
1	Item	Normal Random Value
2	1	54.07640662
3	2	60.58575611
4	3	113.3919968
5		
6	=NORM.INV(RAND(); 75; 15)	

FIGURE 4.11   Excel normals of mean 75 and variance 15.

4.13 GENERATION OF GAMMA RANDOM VARIABLES

Some processes generate positive values skewed to the right. That is, a small proportion of values is observed near the origin; the values become more concentrated, increasing in frequency as they move away from the origin and then their density gradually decreases as they move along the X-axis.

The Gamma random variable can represent continuous values, such as the time it takes to execute an activity; the diameter of trees in a forest; wave height; time intervals between two consecutive failures of a generator; the time it takes for a certain random average λ event to occur β times, volume of rain, service and waiting times, etc. Gamma is a very versatile model when it is required to model a phenomenon that does not admit negative values.

The probability density function Gamma is defined as follows:

$$f(x) = \frac{\beta^{-\alpha} x^{\alpha-1} e^{\frac{-x}{\beta}}}{\Gamma(\alpha)}, \text{ for } x, \alpha, \beta > 0$$

With average $E(x) = \alpha*\beta$, variance $V(x) = \alpha*\beta^2$ and mode $(\alpha-1)*\beta$

The symbol β corresponds to the **scale** parameter and α to the **shape** parameter; although both parameters affect the shape of the distribution. The scale parameter determines the variability of the distribution. The $\Gamma(z)$ is the gamma function defined as:

$$\Gamma(z) = \int_0^\infty t^{z-1} e^{-t} dt \text{ and } \Gamma(n) = (n+1)!$$

Table 4.14 shows some Gamma values.

TABLE 4.14   Gamma values. M. Abramowitz. I. Stegun.

n	1/4	1/2	3/4	2/3	3/2
$\Gamma(n)$	3.625609908	1.77245385	1.2254167	1.3541179	0.88622693

The Excel function =**GAMMALN**(n) can be used to obtain the value of the Gamma function:

$$\Gamma(x) = \int_0^\infty x^{\alpha-1} e^{-x} dx.$$

This is accomplished by entering the value of **x** in the following expression:

=**EXP**(GAMMALN(x)).

For example, **EXP**(GAMMALN(3/4)) = 1.2254167.

To generate Gamma variables the simulation literature offers several procedures, which differ according to the value of the shape parameter. However, Excel incorporates the inverted gamma distribution function, which requires three arguments including a uniform random value and both scale and shape parameters. Excel 2010 offers the function:

=**GAMMA.INV**(RAND();α; β), where α is a shape parameter, and β a scale one.

Example 4.13: Bulb Lifetime

The duration of a light bulb is a Gamma random variable with shape and scale parameters of 3 and 50 days, respectively. Simulate the lifetime of four light bulbs.

Solution
As shown in Figure 4.12 in each cell of the range **B2:B5**, the function =GAMMA.INV (RAND();3;50) is arranged.

	A	B
		Gamma Duration
1	Bulb	(days)
2	1	155.7
3	2	138.4
4	3	38.4
5	4	84.0
6		
7	=GAMMA.INV(RAND(); 3; 50)	

FIGURE 4.12   Bulb duration time.

4.14 RANDOM GAMMA GENERATION: HAHN AND SHAPIRO METHOD

Another alternative to generate Gamma random variables of β scale and of α shape parameters was published by Hahn and Shapiro (1967), according to the following expression:

$$X = -\beta * \left[\sum_{i=1}^{i=\alpha} LN(R_i) \right]$$

Example 4.14: Rainfall
The level of rainfall in a certain region is modeled according to a gamma distribution with parameters of shape $\alpha = 2$ and scale $= 12.5$ mm. Simulate a thousand rains to estimate the amount of rainfall that is above average.

Solution
The shape and scale parameters are arranged in cells A3 and B3, respectively. Applying Shapiro's method requires obtaining the natural logarithm of two random numbers and adding them, and then multiplying the result by the scale parameter (Figure 4.13, cell E6). The contents of cell range A6:E6 are copied to range A7:E1005.

The amount of rain that exceeds 25 liters of water is recorded in cell F6. There is an approximate proportion of 43.2% of rainfall that is expected to exceed 25 liters per square meter.

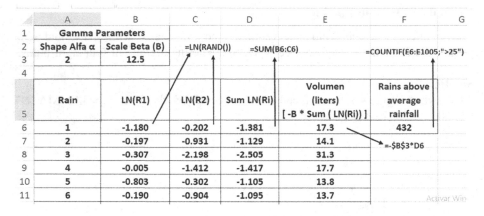

FIGURE 4.13   Simulation of rainfall volume.

4.15 GENERATION OF ERLANG RANDOM VARIABLES

When the shape parameter in a Gamma probability distribution is an integer value, we have an Erlang distribution. Created by Agner K. Erlang to model the number of telephone calls that an operator at a switching station receives at once.

In an Exponential process, if **x** represents the time until one event occurs with media $1/\lambda$, then the convolution $\sum_{i=1}^{i=k} x_i \sim$ Erlang(x, λ, k). That is to say, if **x** is an exponential (λ), then the sum of **k** independent exponential random variables is Erlang (k, λ). Where X_k models the time interval to the **k**th event occurrence. More clearly, **x** variable represents the time interval between any event and the **k**th following event.

The Erlang distribution is useful to model a variety of processes times like time between incoming calls at a Call Center, cell cycle time distributions in medicine, frequency of interpurchase times by customers, time when the **k**th customer arrives at a service, study of wireline telecommunication networks to measure Erlangs: the amount of traffic on telephone systems.

Erlang(x,λ,k) probability density function based on the occurrence rate parameter λ (scale parameter $1/\lambda$) and **k** (shape) parameters is:

$$f(x, k, \lambda) = \frac{\lambda^k * x^{k-1} * e^{-\lambda x}}{(k-1)!}, \text{ where } x, \lambda > 0, \ k = 1, 2, \ldots$$

where **x** represents the time that lapses until the **k**th event occurrence.

Its accumulated distribution function:

$$F(x) = 1 - e^{-\lambda x} * \sum_{n=0}^{n=k-1} \frac{(\lambda x)^n}{n!}$$

F(x) allows us to estimate the probability of the **k**th event occurring before a given time X. That is, P(x<X). With mean and variance:

$$E(x) = \frac{k}{\lambda} \text{ and } V(x) = \frac{k}{\lambda^2}$$

E(x) estimates the average time that the **k**th event will occur.

To generate x ~ Erlang(rate λ, shape k) variable, we consider from probability theory that an Erlang variable is obtained by adding **k** independent random exponential values with mean 1/λ:

$$x = \sum_{n=1}^{n=k} -\frac{1}{\lambda} * LN(R_n)$$

$$x = -\frac{1}{\lambda} * LN\left(\prod_{n=1}^{n=k} R_n\right) \text{ where } R_n \sim U(0, 1)$$

Example 4.15: Device Fidelity and Spare Part

The lifetime of a system component follows an exponential distribution averaging 1,000 hours. If the component fails, an identical replacement component whose life is independent is automatically activated without bringing down the system. Simulate six duration times of the system.

Solution

Figure 4.14 shows the times between failures for the six components in the cell range F5:F10. The main component and its replacement are available. That is, k = 2.

	A	B	C	D	E	F	G	H
1	K	E(x) = 1/λ						
2	2	1000						
3								
4	Component	Random R1	Random R2		R1 * R2	Time to Failure (hours)		
					0.1023	2280.0	→ -$B$2*LN(E5)	
5	1	0.3882	0.2635				→ B5*C5)	
6	2	0.5165	0.4867		0.2514	1380.7		
7	3	0.2015	0.9640		0.1943	1638.6		
8	4	0.0502	0.7869		0.0395	3230.7		
9	5	0.7495	0.5294		0.3968	924.4		
10	6	0.1558	0.4063		0.0633	2759.6		
11								
12		=RAND()						

FIGURE 4.14   Life time of component.

The duration of the system until it fails is an Erlang random variable of order 2 and rate $\lambda = 0.001$.

4.16 GENERATION OF BETA RANDOM VARIABLES

The Beta distribution is useful to represent processes whose events can be measured as values which correspond to proportions, percentages, or fractions; for example, daily percentage of damaged machines, monthly proportion of sick workers, percentage of companies that close annually, fraction of students passing each semester. It can also be used to model activity duration times. For random between minimum and maximum limits [a,b], with shape parameters $\alpha1$ and $\alpha2$, and Beta function $B(\alpha1, \alpha2)$ its density function is:

$$f(x) = \frac{x^{\alpha1-1}(1-x)^{\alpha2-1}}{B(\alpha1, \alpha2)}$$

A generating method, according to Banks (1998): If $G_1(a, \alpha1)$ and $G_2(a, \alpha2)$ are two Gamma random variables of parameters $(a, \alpha1)$ and $(a, \alpha2)$, respectively, then

$$X = \frac{G_1}{G_1 + G_2}$$

It follows a Beta distribution of shape parameters $\alpha1$ and $\alpha2$.

Example 4.16: Revenue from Maintenance Services
The percentage of weekly revenue coming from maintenance services in the workshop of a car dealership is random Beta with parameters 5 and 4. Generate 6 weeks of the percentage of revenue coming from the workshop (Figure 4.15).

	A	B	C	D	E	F
		Random Gamma	Random Gamma	% of Revenue		
1	Week	G(2,5)	G(2,4)	(Beta)		
2	1	3.37	7.44	31.1	→=(B2/(B2+C2))*100	
3	2	9.98	4.56	68.6		
4	3	7.69	6.52	54.1		
5	4	5.92	2.81	67.8	↓	
6	5	6.73	8.48	44.3		
7	6	16.82	3.75	81.8		
8			↓			
9			=GAMMA.INV(RAND(); 2; 4)			
10	=GAMMA.INV(RAND(); 2; 5)					

FIGURE 4.15   Percentage of weekly revenue.

4.16.1 Random Beta with Excel Function

A more direct alternative for Beta generation is through the Excel function:

=**BETA.INV**(RAND(); 5; 4)

	A	B
		% of Revenue
1	Week	(Beta)
2	1	0.6879
3	2	0.4527
4	3	0.5216
5	4	0.4662
6	5	0.5140
7	6	0.5642
8		
9	=BETA.INV(RAND(); 5; 4)	

FIGURE 4.16   Beta values (5;4) with Excel.

4.17 WEIBULL RANDOM GENERATION

In 1939, the Swedish physicist Waladdi Weibull introduced the probability distribution that bears his name. This distribution is a powerful analysis tool for symbolizing material breakage resistance, reliability, and lifetime of entities.

It has been used to model strike duration, AIDS mortality, survival time of entities subject to failure, duration of earthquakes, time to perform an activity, diameter of trees, duration of marriages, amount of rain precipitation, packet inter-arrival time in a LAN or local area network, etc. Its probability density function is as follows:

$$f(x) = \frac{\beta}{\alpha}\left(\frac{x}{\alpha}\right)^{\beta-1} e^{-\left(\frac{x}{\alpha}\right)^{\beta}}, \text{ for } x \geq 0$$

With its scale $\alpha > 0$ and shape $\beta > 0$ parameters. When it has three parameters, it includes a location parameter, which can be interpreted as its minimum value. The scale parameter α relates to the path of the x-values of the distribution on the abscissa axis. α expresses the units of its variable. The shape parameter β represents the failure rate, which can be constant, increasing, or decreasing and reflects the shape of the distribution.

If a Weibull random event collects the time duration until a failure occurs, the failure rate β can be interpreted as follows (Table 4.15):

TABLE 4.15   Interpretation of the failure rate

$\beta < 1$	Failure decreases over time	Infant adaptation
$\beta = 1$	Indicates constant failure over time	Cause: External events
$\beta > 1$	Failure increases with time	Aging process

It has averaging or mean time between failures as follows:

$$E(x) = \frac{\alpha}{\beta} \Gamma\left(\frac{1}{\beta}\right)$$

and variance:

$$V(x) = \frac{\alpha^2}{\beta}\left\{2\Gamma\left(\frac{2}{\beta}\right) - \frac{1}{\beta}\left[\Gamma\left(\frac{1}{\beta}\right)\right]^2\right\}$$

Its cumulative probability function:

$$F(X) = 1 - e^{-\left(\frac{x}{\alpha}\right)^{\beta}}$$

Like the uniform and exponential probability functions, the Weibull is a continuous strictly increasing distribution, easy to obtain its inverse. Consequently, to find a random generator algorithm, the inverse transform method described previously in Table 4.5 can be applied, from which the following Weibull generator procedure is obtained:

$$x = \alpha * [-\ln(R)]^{\frac{1}{\beta}} \tag{4.2}$$

where R is a uniform random value between 0 and 1.

The 63rd percentile of the Weibull distribution is approximately the value of the scale parameter α. It can be calculated with:

$$Y_p = \alpha[-\ln(1 - p)]^{\frac{1}{\beta}}$$

Example 4.17: Generate Weibull values for a Weibull variable with Beta = 2 and Alpha = 13. Verify that the 63rd percentile is approximately 13. Also, determine that the 90th percentile corresponds to 19.73.

Solution

In Figure 4.17, Cell B7 shows that 63.21% of the values are below the scale parameter 13 of the Weibull(2; 13). Similarly, cell E4 shows that about 90% of the values are less than 19.73.

	A	B	C	D	E
1	Y	Probability(x < Y)		α	13
2	8	0.3152		β	2
3	9	0.3808		% of Percentile =	90
4	10	0.4466		Percentile =	19.727
5	11	0.5113			
6	12	0.5735		=E1*(-LN(1-E3/100)^(1/E2))	
7	13	0.6321			
8	14	0.6864			
9	15	0.7359			
10					
11	=WEIBULL.DIST(A2; 2; 13; 1)				

FIGURE 4.17   Obtaining the 63rd percentile for the Weibull.

4.17.1 Random Weibull with Excel

To obtain random Weibull values, the procedure from the application of the inverse transform method presented in expression 4.2 is used.

To date Excel does not offer a random Weibull generator. However, it does have a function to calculate its probabilities. Its syntax is as follows:

=WEIBULL.DIST(x; shape parameter β; scale parameter α; 1 if accumulated).

Example 4.18: Lifetime of a Tablet
The time until a computer tablet fails is Weibull of scale parameters 500 days and shape 5. Simulate the lifetime of five digital devices.

Solution
In each of the cells of the range B5:B9 (Figure 4.18), the generator for Weibull random values is written directly as:

	A	B	C	D	E
1	Scale =	α =	500		
2	Shape =	β =	5		
3					
4	Tablet	Duration (Days)			
5	1	614.0	=$C$1*(-LN(RAND())^(1/$C$2))		
6	2	444.6			
7	3	447.3			
8	4	681.8			
9	5	438.0			

FIGURE 4.18   Generation of Weibull random values.

= 500*(-LN(RAND())^(1/5)).

It is also possible to refer both parameters in independent cells. The scale parameter 500 is in cell **C1**, and the shape parameter in cell **C2**.

Example 4.19: Duration of an Electric Fence Equipment
The operating time of an electric fence equipment follows a Weibull probability distribution with 10,000 hours and failure rate 1.2. (A) What proportion of equipment is expected to last more than 10,000 hours? (B) Obtain the mean time between failures of a piece of equipment. (C) Simulate to verify answers (A) and (B).

Solution

A. P(X >10,000) = 1-**WEIBULL.DIST**(10000;1.2; 10000;1) = 0.3679
About 37% of the equipment is expected to operate more than 10,000 hours.
B. The mean time between failures is achieved according to:

$$E(x) = \frac{\alpha}{\beta} \Gamma \left(\frac{1}{\beta} \right)$$

= (10000/1.2)*EXP(GAMMA.LN(0.83333)) = 9,406.6 hours.

The average life time of the equipment is more than 9,406 hours.

C. In Figure 4.19, the Excel expression:

=10000*(-LN(RAND())^(1/1.2))

Arranged in cells **B2:B10001**, allows to generate 10,000 Weibull random numbers, a simulation which mimic the lifetime of 10,000 electric fence energizing equipment. Cell **D2** contains the value of the expected duration time, whose simulated result is 9,367.7 hours. Cell **F2** shows the probability that a fence lasts more than 10,000 hours (Figure 4.19).

	A	B	C	D	E	F
		Duration		Simulated Average Life	Equipments lasting >	Probability Duration >
1	Equipment	(Hours)		Time (hours)	10000 hours	10000 hours
2	1	9,162.4		9,367.7	3650	0.365
3	2	766.8				
4	3	7,297.6		=AVERAGE(B2:B10001)		=E2/10000
5	4	1,924.1				
6	5	505.2		=COUNTIF(B2:B10001; ">10000")		
7	6	11,383.2				
10000	9999	10,577.4				
10001	10000	8,512.7		=10000*(-LN(RAND())^(1/1.2))		

FIGURE 4.19 Weibull fence equipment duration.

4.18 GENERATION OF LOGNORMAL RANDOM VARIABLES

Certain processes generate response values originated from the accumulation of several independent random factors. The added response of such processes can be modeled according to a normal probability distribution. Similarly, if the outcome of a random event is a consequence of the multiplicative product of several independent factors or effects, then it is possible to represent its occurrence as a lognormal random variable. Also, when the mean of the data is low, with wide variance, pronounced skewness and non-negative values, the lognormal distribution could be useful.

This relative of the normal distribution is applicable to symbolize various biological, physical, social, and economic processes. It can be used to model events whose range of values covers powers of 10. For example, activity duration times; time it takes for a sick person to recover; annual electricity consumption of subscribers; stock market share price; vitamin B12 concentration in the blood; duration of earthquakes; latency time of diseases from contagion to onset of symptoms; particle diameter after crushing; financial loss; reliability of components; distribution of chemicals and organisms in the environment; length of time of marriage of couples; survival time after diagnosis of cancer or AIDS; number of animal species per family; age of onset of Alzheimer's disease (60 years; 1.16) (Horner 1987).

4.18.1 Lognormal Distribution

If x is a normal random variable of mean μ and standard deviation σ, then variable $Y = e^x$ has Lognormal distribution of lognormal mea μ_L and lognormal standard deviation σ_L. That is, a random variable x is said to be lognormally distributed if **Log** (**x**) is distributed according to a normal distribution. This distribution is skewed to the left, more dense and only admits positive values. The Lognormal probability distribution has the following expression:

$$f(x) = \frac{1}{x\sqrt{2\pi}\sigma^2}e^{\frac{-(\ln(x)-\mu)^2}{2\sigma^2}}, \quad Si \ x > 0$$

With lognormal mean μ_L:

$$E(x) = e^{\mu + \frac{\sigma^2}{2}}$$

Lognormal variance:

$$V(x) = e^{2\mu + \sigma^2} * \left(e^{\sigma^2} - 1\right)$$

If a random variable is lognormal, its logarithm can be extracted and normal values can be obtained, to which the procedures requiring data from a normal distribution can be applied.

4.18.2 Lognormal Generation Procedure

To generate lognormal random values whose lognormal mean is μ_L and lognormal deviation σ_L, the following procedure is followed.

TABLE 4.16 Procedure for generating lognormal random variables

1. Knowing the parameters of a normal distribution: mean μ and deviation σ. Use the parameters of the Lognormal distribution according to the following expressions that relate the parameters of both normal and Lognormal distributions:

$$\mu = LN\left(\frac{\mu_L^2}{\sqrt{\mu_L^2 + \sigma_L^2}}\right)$$

$$\sigma = \sqrt{LN\left(\frac{\mu_L^2 + \sigma_L^2}{\mu_L^2}\right)}$$

2. Achieve a normal random value of mean μ and standard deviation σ by any normal random generating algorithm, e.g., from the Excel expression::

$$x = \text{NORM.INV(RAND(); } \mu; \sigma)$$

3. Get the lognormal random value Y, from $Y = e^x$
4. Repeat steps **2** and **3** for each lognormal random value required

4.18.3 Lognormal Generation in Excel

The following Excel expression directly generates Lognormal random values:

=LOGNORM.INV(RAND(); μ normal; σ normal)

Figure 4.20, the range of cells F5:F14, presents 10 values generated with the inverted Lognormal Excel function. Note that the arguments included in the function correspond to the parameters obtained after using the equations for μ and σ calculated in step 1 of the procedure described in Table 4.16.

LogNormal Mean= 10		Mean m = 2.259 ← =LN(100/SQRT(109))	
LogNormal Deviation = 3		Deviation d = 0.294 ← =SQRT(LN(109/100))	

Device	A) Duration (years) LogNormal Random X, using =LOGNORM.INV(RAND();2.259; 0.294)	B) LN Random X (years) Exponentiating a Normal(2.259, 0.294)	C) Most convenient method Using =LN(LOGNORM.INV(RAND();LNMean;LNDev)) (Years)
1	15.38	13.89	11.96
2	11.77	9.04	14.30
3	9.97	10.59	13.22
4	9.66	6.43	9.79
5	10.34	5.08	7.63
	=LOGNORM.INV(RAND();$E$1;$E$2)	=EXP(NORM.INV(RAND();$E$1;$E$2)))	=LN(LOGNORM.INV(RAND();$B$1;$B$2))

FIGURE 4.20   Three ways to generate random Lognormals.

Example 4.20: Lifetime of Electronic Components

The lifetime of an electronic device conforms to a lognormal distribution with average 10 years and deviation equal to 3. Simulate the lifetime of five devices. (A) Using the procedure given in Table 4.16. (B) Most directly, with the inverse lognormal function of Excel.

Solution

A. Calculate the normal mean and deviation to be used in the lognormal generation algorithm:

$$\mu = LN\left(\frac{10^2}{\sqrt{10^2 + 3^2}}\right) = 2.26$$

$$\sigma = \sqrt{LN\left(\frac{10^2 + 3^2}{10^2}\right)} = 0.2935$$

Generate a normal random **x** value of mean 2.26 and standard deviation 0.2935 and calculate **Y** = Exp(x) (Figure 4:20, columns B and D). This procedure is subject to determining from the mean and lognormal deviation, the parameters m and d. The previous procedure requires additional calculation effort.

A more immediate option is presented in column D. It only requires embedding the desired mean and lognormal deviation in the Excel generator and obtaining its natural logarithm. The generation of lognormal with the Excel function is shown in Figure 4.20, column E, range E5:E9.

Figure 4.21 presents lognormal values using a simplest procedure. Through both lognormal parameters in Excel's LOGNORM.INV() function, and then extracting the logarithm to get the duration of each electronic device.

It can be verified that the average and the deviation of the generated lognormal values approximate the expected values 10 and 3 years lasting time.

	A	B	C	D
		Random	Duration	
1	Device	Lognormal	(years)	
2	1	12,789.40	9.5	=LN(B2)
3	2	516,825.56	13.2	
4	3	2,797.20	7.9	
5	4	25,180.81	10.1	
6	5	748.42	6.6	
7				
8	LOGNORM.INV(RAND(); 10; 3)			

FIGURE 4.21   Lognormal variates.

4.19 GENERATION OF TRIANGULAR RANDOM VARIABLES

The triangular probability distribution of parameters minimum **a**, most probable **b** and maximum **c**, is conveniently used when data on the event to be represented are not available, so that in the simulation of an event approximated values or educated guess can be assumed for its three parameters.

Its density function has the following expression:

$$f(x) = \frac{2(c - x)}{(c - a)(c - b)}, \quad \text{if } b \le x \le c$$

$$f(x) = \frac{2(x - a)}{(b - a)(c - a)}, \quad \text{if } a \le x \le b$$

Average:

$$E(x) = \frac{a + b + c}{3}$$

And variance:

$$V(x) = \frac{a^2 + b^2 + c^2 - ab - ac - bc}{18}$$

4.19.1 Triangular Generation Procedure

Algorithm 1
Banks (1998, p. 159) exposes the following Triangular random generation procedure (Table 4.17):

TABLE 4.17   Triangular random generation

Giving parameters a, b, and c
$$Z = (b - a) / (c - a)$$
Generate a uniform random value **R** between [0,1)
If **R < Z**, calculate:
$$T = \sqrt{Z * R}.$$
otherwise:
$$T = 1 - \sqrt{(1 - Z) * (1 - R)}.$$
$$x = a + (c - a) * T$$
x is a triangular random variable

Example 4.21 Duration of a Job

The time it takes a heavy equipment operator to complete a job is uncertain. It is known that it will never take less than 2 hours, not more than 7; although the most likely time is 4.5 hours. Simulate the duration time for five jobs.

Solution

The input data are: **a** = 2, **b** = 4.5, and **c** = 7. After substituting the input values in the generation procedure, we have the triangular random values shown in column D5:D9 (Figure 4.22).

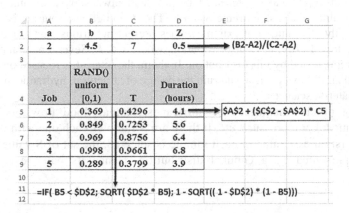

FIGURE 4.22   Duration time for five jobs.

An **alternative** more direct **procedure to obtain triangular random values** is the following:

$$= c + (a + RAND() * (b - a) - c) * SQRT(RAND())$$

See Figure 4.23. By example, with **a** = 2, **b** = 4.5, and **c** = 7.

	A	B	C	D	E
1	**a**	**b**	**c**		
2	2	4.5	7		
3					
4	**Job**	**Duration (hours)**			
5	1	3.7			
6	2	4.1			
7	3	2.2			
8	4	4.2			
9	5	3.9			
10					
11	=$C$2 + ($A$2 + RAND() * ($B$2 - $A$2) -$C$2) * SQRT(RAND())				

FIGURE 4.23 A simpler procedure for triangular generation.

4.20 GENERATION OF GUMBEL OR EXTREME VALUE RANDOM VARIABLES

A series of different values, repeated or not, will always have a minimum and a maximum value, that is, extreme values. The Gumbel distribution is useful for representing the distribution of extreme values present in various data sequences. Here refers to the type I extreme value distribution; where $x \in R$. Several natural phenomena have been represented by this distribution. In them, the interest is focused on evaluating the values that, if they occur, would ruin the design of structural, hydraulic, or resistance systems, material sciences.

Among others, magnitude of hurricanes, annual flood flows, earthquakes, rainfall, atmospheric pressure, etc. If there are **S** samples each one with **n** values, if the largest (smallest) value of each sample is selected, the set of **S** values obtained can be represented by a Gumbel distribution. Hence its name extreme value function.

4.20.1 Probability Density Function of parameters μ and β

$$f(x) = \frac{1}{\beta} * e^{-\left(\frac{x-\mu}{\beta}\right)} * e^{-e^{-\left(\frac{x-\mu}{\beta}\right)}}, \ -\infty < x < \infty$$

4.20.2 Getting Gumbel Parameters

Obtain the average of the values to estimate the parameters from:

Scale: $\beta = \frac{\bar{x}}{1.28255}$ and Location: $\mu = \bar{x} - 0.577216 * \beta$

where $\bar{x}$ is the mean of the data.

4.20.3 Descriptive Statistics

Mean: $E(x) = \mu + 0.577216 * \beta$
Variance: $V(x) = 1.644934 * \beta^2$
Median: $\mu\text{-ln(ln(2))} * \beta$
Mode: μ

4.20.4 Cumulative Probability Distribution

Gumbel cumulative probability distribution: the probability that a value less than x will occur:

$$F(x) = e^{-e^{-\left(\frac{x-\mu}{\beta}\right)}}$$

4.20.5 Gumbel Random Variates

The random Gumbel values x are generated by the following relationship:

$$x = \mu - \beta * \text{LN}(-\text{LN}(R))$$

where μ relates to the location parameter and β to the scale parameter. R corresponds to a uniform random value between 0 and 1.

Example 4.22: Generate 90 random Gumbel values of parameters location 82 and scale 25. Obtain their theoretical and simulated average.

	A	B	C	D	E	F
1	Mu (location)	Beta (scale)			E(x) Theoretical	E(x) Simulated
2	82	25	=A2+0.5772*B2 ────		→ 96.43	96.6
3	=$A$2-$B$2*LN(-LN(RAND()))				=AVERAGE(A5:F19)	
4						
5	84.9	113.3	145.3	94.7	203.1	76.5
6	114.1	80.8	70.4	72.8	100.8	154.3
7	73.5	113.7	63.7	99.4	134.2	104.1
8	90.8	60.5	82.9	130.5	76.8	75.8
9	61.3	61.8	72.4	110.3	112.2	93.9
10	102.2	42.3	110.8	91.4	76.0	47.4
11	89.4	105.1	118.7	81.6	47.2	114.9
12	71.4	108.8	75.4	92.3	87.8	195.5
13	198.9	89.1	92.2	156.7	103.3	87.0
14	83.3	221.4	69.7	65.9	67.7	59.0
15	71.0	77.1	133.3	89.4	89.4	94.5
16	85.8	69.6	97.1	119.3	69.1	113.3
17	102.2	125.4	76.8	97.9	91.5	29.7
18	120.9	51.1	95.4	96.5	61.8	100.5
19	94.7	146.5	72.0	131.6	145.8	63.1

FIGURE 4.24   Ninety gumbel variates.

Solution

In Figure 4.24, all cells in the range A5:F19 contain the formula:

=$A$2-$B$2*LN(-LN(RAND(())))

which generates 90 Gumbel values. The expected and simulated averages are practically similar (Figure 4.24, cells E2 and F2).

Design of Simulation Experiments

5

5.1 SIMULATION LENGTH: HOW MANY RUNS?

Although the realization of a simulation study corresponds to an experimental test on a surrogate model of reality, the scientific method of observation and experimentation is absolutely applicable to the model under analysis. The reasoning on the statistical results of the simulated experiment is inductive, although under a certain level of uncertainty outcomes are originated from the probabilistic sampling of a model. Reality and simulation models are full of data and uncertainty.

The simulation method performs a controlled imitation experiment of systems whose behavior is generally governed by chance. To achieve reliable results, the simulation approach requires determining the appropriate sample size or length of sampling time, which allows estimating from a sample the population parameters with reliable accuracy.

The fundamental tools for assessing chance and its instability are probability models. The statistical law of large numbers guarantees that the average of values of a series of n trials converges to the expected value of the parameter being estimated. Moreover, the accuracy of the estimation increases as the sample **n** becomes larger; although after a certain number of simulations, the improvement slows down and does not pay off by running additional simulations.

The size of a statistically representative sample of a population depends on several factors, such as homogeneity or variability of the elements that make up the population, confidence in the estimate obtained, desired accuracy, distribution of interest variable, correlation of the elements of the population, size of the population, whether finite or infinite, and availability of the population variance. In the end, the sampling theory justifies its application in the scarcity of resources to census all the entities of the population. The simulation experiments referred to in this book assume infinite populations, with unknown variance and large samples, from statistics view

DOI: 10.1201/9781032701554-5

point larger than 30 observations. This reduces the formal statistical effort to determine the length of the simulation to simpler expressions.

In general, a simulation study may be aimed at estimating population parameters that represent:

1. Proportions, probabilities, percentages, fractions.
2. Average values.

The sampling procedures to be used require defining aspects indicated in Table 5.1.

TABLE 5.1   Requirements for a sampling process

1. Confidence level $(1 - \alpha)$ to be used in the estimation.
2. Accuracy or maximum error $(\in)$ to be tolerated due to the difference between the value of the parameter μ and the value of the estimated statistic $\bar{x}$.
3. Variance (σ^2) of the values under study.

In case of the estimation of averages, the variability of the values is collected in the sampling variance s^2, calculated from a pilot run. Since in practice, computer sampling does not generate costs, a pilot sampling of size between 500 and 1,000 values of the variable of interest can be carried out. This helps to more accurately estimate the variance of the variable and improve its contribution to the formula used to determine the simulation length.

Running a simulation model containing chance behavioral factors only once delivers a random result from a wide range of possible responses: Random inputs generate random responses. That is, each simulation run gives a different answer, subject to unplanned variability which requires a statistical analysis of the results.

In general, if we have **n** observations, $X_1, X_2, X_3, \ldots, X_n$, from a sample to estimate a parameter corresponding to a population average μ, we obtain a minimum variability estimator $\bar{x}$, unbiased and consistent, by means of:

$$\bar{x} = \frac{1}{n} \sum_{i=1}^{i=n} x_i$$

That is, to determine the expected value of a random variable defined as a population parameter μ, a statistic $\bar{x}$ is estimated from a random pilot sample. The error of the estimate $|\mu - \bar{X}|$ is the difference between the value of the population parameter μ and its sample statistic $\bar{x}$.

Chebyshev's theorem allows us to establish:

$$P(\mu - \bar{X}) < 3 * \frac{\sigma}{\sqrt{n}} = 0.998$$

That is, 99.8% of the values of a probability distribution are expected to be within three deviations of its mean.

A confidence interval of $(1 - \alpha)\%$ is obtained by means of:

$$\bar{x} \pm t_{n-1,\ 1-\alpha/2} * \frac{s}{\sqrt{n}}$$

5.2 SIZE n: PROPORTION OR POPULATION PROBABILITY p

The sample size for proportions or probabilities requires as input information the confidence level $1 - \alpha$, expressed between 0 and 1, and the tolerable error $\in$. The length of the simulation is determined according to the following relation:

$$n \geq \frac{Z_{\alpha/2}^2 * p(1 - p)}{\in^2} \tag{5.1}$$

where $Z_{\alpha/2}$ represents the corresponding Z value of a standard normal distribution; and $\in$ to the desired error or accuracy, expressed between 0 and 1. The p value can be estimated without performing a pilot run. By definition p is a value between 0 and 1. If a value of p less than or greater than 0.5 is assumed, the product p*p would be less than 0.25. So p = 0.5 is kind to the sample size. For this purpose, a favorable ratio $p = 0.5$ can be assumed. In this case,

$$p * (1 - p) = 0.5 * (1 - 0.5) = 0.25 = 1/4$$

which simplifies expression 5.1 to:

$$n \geq \frac{Z_{\alpha/2}^2}{4 * \in^2}$$

Example 5.1: Dice: Number of Throws
Three dice signed D_1, D_2, and D_3 are thrown twice. Determine the sample size to estimate the probability of obtaining the same result in both throws, for a confidence level of 99%, and a maximum error of 0.5%.

Solution
For: (A) $1 - \alpha = 0.99$. (B) $\alpha = 0.01$. (C) $\in = 0.005$. (D) $Z_{\alpha/2} = Z_{0.005} = 2.5758$.
    The Excel sheet gives the standard normal value by: =**NORM.S.INV**(0.005)

$$n \geq \frac{Z_{0.005}^2}{4 * 0.005^2} = \frac{6.6348966}{0.0004}$$

$$= 66,348.9 = 66,349 \text{ throws}$$

That is, to estimate the probability $\hat{p}$ of achieving the same result when performing both throws, with 0.5% precision and 99% confidence, it is required to perform a simulation of length greater than 66,000 throws.

5.3 SIZE n: ESTIMATION OF A POPULATION AVERAGE μ

If the population parameter to be estimated is an average, a consequence of the accumulation of a series of values, it is necessary to determine whether the behavior of the variable of interest can be assumed to be distributed according to a normal random variable.

This requires simulating a pilot sample that produces a sequence of values of the variable of interest, to assess whether the data could be assumed to be approximated normal.

Any of several statistical procedures aimed at testing hypotheses of normality can be used. To evaluate visually, it is recommended to obtain about a thousand values to construct a histogram of the variable and to observe if the shape of the values to be averaged is bell-shaped, which would suggest a Gaussian distribution. Additionally, a descriptive statistical analysis is performed to corroborate if the mean, median and mode are approximately equal; and to make sure if the kurtosis and skewness are close to 0. To decide on the procedure to determine the length of the simulation and to estimate an average, one of the following options should be evaluated and chosen:

1. The values are distributed according to a normal function and therefore the Central Limit Theorem (CLT) is applicable.
2. Normality of the variable of interest cannot be assumed. Consequently, Chebyshev's inequality will be used to determine the sample size.

Option 1. If CLT is applicable, the simulation length can be achieved from the following relationship:

$$n \geq \frac{Z_{\alpha/2}^2 * S^2}{\in^2}$$

where S^2 corresponds to the variance of the pilot sample. Likewise, $Z_{\alpha/2}$ is the corresponding Z value from a standard normal distribution, with significance level α. In addition, $\in$ is the desired error or accuracy. $\in$ must be expressed in the same units of the variable to be estimated.

Option 2. If the values obtained from the pilot sample do not show a tendency to a normal distribution, although it will deliver a larger sample size, the following relationship based on Chebyshev's theorem can be employed:

$$n \geq \frac{S^2}{\alpha * \in^2}$$

where S^2 corresponds to the variance calculated with the pilot sample values. Likewise, α is the significance level and $\in$ is the desired error or accuracy.

Example 5.2: Truck Production

The number of trucks produced daily on an assembly line follows the following empirical probability distribution (Table 5.2):

TABLE 5.2 Number and probability of trucks produced daily

Trucks	195	196	197	198	199	200	201	202	203	204	205	206
Prob.	0.03	0.05	0.07	0.10	0.15	0.20	0.15	0.10	0.07	0.05	0.02	0.01

At the end of the day the finished trucks are moved on a shuttle through a bay. However, the capacity of the ferry is 198 trucks. When the day's production exceeds that capacity, the remaining trucks are moved on external cargo ships, which charge $50,000 per truck moved. Determine the length of the simulation to estimate, with 95% confidence and an error of $5,000, the average daily cost expected to be paid by the assembler to the external vessels.

Solution

The expected daily payment cancelled for the use of external vessels corresponds to the average of the disbursements made daily. It is required to simulate a number of days using a pilot sample. Then evaluate the possibility that the values of the random variable of interest: daily payment, are distributed according to a normal probability function. This allows us to decide whether to apply the CLT or Chebyshev's theorem. Figure 5.1 presents the first 4 days of simulation of the pilot run for 1,000 days.

In this case, a preliminary sample of 1,000 days is taken to obtain 1,000 values of daily payments for truck loading on external vessels. The daily payment amounts from this pilot sample can be evaluated to determine if they can be assumed to be representative of a normal distribution. An exploratory method is to take the values to construct a frequency histogram and observe whether the histogram can be assumed to be equivalent to that of a normal distribution.

When drawing the histogram, it can be seen that the graph does not resemble a normal shape (Figure 5.3). Excel allows you to draw histograms with its **Data Analysis** module, which can be activated from the toolbar, **Data** option. In addition,

	A	B	C	D	E	F	G	H	I	J	K	L	M
1	Daily Production:	195	196	197	198	199	200	201	202	203	204	205	206
2	Probability:	0.03	0.05	0.07	0.1	0.15	0.2	0.15	0.1	0.07	0.05	0.02	0
3	0	0.03	0.08	0.15	0.25	0.4	0.6	0.75	0.85	0.92	0.97	0.99	1
4													
5	Ferry Capacity:	199	=A3+B2										
6	External Cost $/truck:	50,000.00	=LOOKUP(B9; $A$3:$M$3; $B$1:$M$1)										
7													
8	Day	R	Truck Daily Production	Trucks trasladed by external Vessels	Daily payments to external vessels	=IF(C9 <=$B$5; 0; C9-$B$5)							
9	1	0.765	202	3	150,000.00	=D9*$B$6							
10	2	0.792	202	3	150,000.00								
11	3	0.867	203	4	200,000.00								
12	4	0.593	200	1	50,000.00								

FIGURE 5.1   Simulation of production and truck transfer.

the **Descriptive Statistics** option is available in **Data**, which provides information on the characteristic measures of any set of values. If a data series tends to a normal distribution, its median, mean, and mode should resemble each other. In addition, the skewness and kurtosis coefficients (under Excel) should approach 0.

Therefore, in this case (Figure 5.2), the daily payments for external vessel usage do not meet the normality condition. Consequently, the sample size determination should be based on the Chebyshev method.

The sample variance of daily payments calculated from the pilot sample of 1,000 values yields a value of 7,670,848,348.3. In order to have 95% confidence, and an estimate of the average daily payment with error less than $5,000, the length of the simulation is obtained by applying the Chebyshev relation, being necessary to simulate the process for at least 6,137 days.

	A	B
1 2	**Daily payments to external vessels**	
3	Mean	76,350
4	Standard Error	2,770
5	Median	50,000
6	Mode	0
7	Standard Deviation	87,583.4
8	Sample Variance	7670848348.3
9	Kurtosis	0.50941
10	Skewness	1.112
11	Range	350,000
12	Minimum	0
13	Maximum	350,000
14	Sum	76350000
15	Count	1000

FIGURE 5.2   Daily payment statistics.

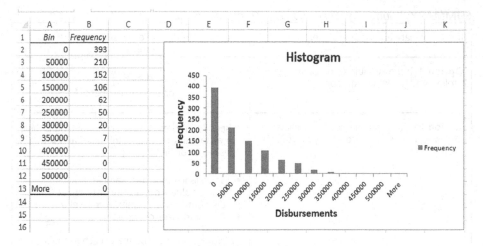

FIGURE 5.3 Distribution of daily payments.

$$n \geq \frac{S^2}{\alpha * \in^2} = \frac{7,670,848,348.3}{0.05 * 25,000,000}$$

$$n = 6,137 \; dias$$

In general, when more than one parameter needs to be estimated in a simulation study, the sample size of each parameter should be determined and then the largest value from among the sample sizes obtained separately should be chosen as the final size.

5.4 EMPIRICAL SIZE OF THE SIMULATION

A practical method to determine the number of simulations to perform is to run two independent repetitions, each of **n** observations, to obtain two results **R₁** and **R₂** and compare both responses. If both results are different, double the number **n** of simulations to **2∗n** and recalculate two new results. This procedure is repeated, doubling the previous number of simulations until the difference or error between the two results is considered minimal and satisfactory.

For example, start with **n** = 5,000; then **n** = 10,000; 20,000, … . This empirical method constitutes a functional albeit informal alternative to achieve a simulation length without statistical rigor. The method allows to approximate the parameter of interest by approaching its average, decreasing the divergence between the population value (parameter) and its estimated value (statistic) as the simulation length grows.

SUMMARY: *Summary: Simulation Length or Sample Size n*

If you estimate an Average value:	If you estimate a Percentage, Proportion, or Probability:
Option 1. If the variable to be estimated follows a Normal distribution. $$n > =\frac{Z_{\alpha/2}^2 \cdot S^2}{E^2}$$	$$n > =\frac{Z_{\alpha/2}^2}{4*E^2}$$ Error E between 0 and 1
Option 2. If the variable does not follow a Normal distribution. Use Chebyshev's formula: $$n > =\frac{S^2}{\alpha *E^2}$$	Where $Z_{\alpha/2}$ represents the corresponding Z value of a standard normal distribution; and **E** the desired error or accuracy. • S^2 variance of a pilot sample. • $Z_{\alpha/2}$ Z value from a standard normal distribution with significance level α. • E desired error or accuracy. Warning: The error units E in each formula must be expressed according to the type of variable to be estimated. If average type: E value must be in the same units of the variable to be estimated. Example 560. If probability, percentage, or proportion: E value should be between 0 and 1. Example 0.07.

5.5 INITIAL CONDITIONS FOR SIMULATION

The simulation experiment starts the model at time 0 and at a certain starting state. This initial condition will influence the performance of the model for a certain period of time, decreasing its impact as the simulation time progresses. When the result of the parameter to be estimated is independent of the initial conditions, we have a steady state or long term estimation. This is called steady-state simulation. It is called transient-state simulation when the initial conditions significantly affect the estimation of the parameter of interest.

Generally, long-term behaviors are simulated, where the starting values are diluted over the observation time. It is recommended to start the model with adherence to the usual conditions in which the real system starts.

5.6 CONSTRUCTION OF CONFIDENCE INTERVALS

The simulation process requires the use of basic statistical tools to analyze the results generated by stochastic events. It is important to understand the random variability of the results in each simulation run. The output delivered by the model run is only one value out of an infinite number of possible outcomes. It is an experiment with a lot of variability and randomness. Since the values of the run are random and at each opportunity it offers different results, these responses must be characterized by a sampling distribution, which describes all possible results of the samples of the system or population under study. It is convenient to emphasize the difference between a population parameter and its sample estimator called statistic.

The parameter μ is a constant and unknown value. The statistic $\overline{X}$ is a random variable originated by random sampling, which takes a different value in each run. Therefore, in addition to determining the length of a simulation run, it is necessary to perform n independent repetitions. A confidence interval expresses the uncertainty in an estimate. Due to chance, an interval constructed from one sample is different from that obtained from another sample.

To refine the estimate several uncorrelated runs or repetitions should be executed, so that in each run an independent estimate of the value of interest is obtained. For example, twenty runs and average the results. The data thus obtained correspond to random estimates of the parameter, evaluated from n independent samples.

Their average and standard sample deviation allow defining the confidence interval at the desired level. The values, each one coming from the average over the results obtained in each simulation, faithfully comply with the powerful CLT: as n increases, the sample distribution of the statistic $\bar{x}$ approximates a normal distribution with variation or standard error: $\frac{s}{\sqrt{n}}$.

According to the CLT, regardless of the type of data, the distribution of the sum of the averages of a series of n random values follows a normal distribution. That is, if several random samples of equal size are taken from the same population, the histogram of the averages will show a normally distributed trend. This allows to take A_i averages obtained from n independent runs, although empirically, it can be used between 10 to 30 runs.

Obtain their average and determine the sampling deviation s and construct a confidence interval of $(1 - \alpha)$ % by the expression:

$$\bar{x} \pm t_{1-\frac{\alpha}{2};\, n-1} * s/\sqrt{n}$$

$$\bar{x} = \frac{1}{n} \sum_{i=1}^{i=n} A_i$$

$$s^2 = \frac{1}{n-1} \sum_{i=1}^{i=n} (A_i - \bar{x})^2$$

A confidence interval defines a range of values, minimum and maximum. The $(1 - \alpha)\%$ of the time may contain the value of the parameter μ to be estimated. In general, simulation is performed to estimate the average performance of a system. It is essential to accompany the estimate of the unknown parameter μ with its respective confidence interval.

Example 5.3: Waiting Times

Ten repetitions of the simulation of a service process yielded the following estimates for the waiting time in minutes:

12.6	15	12.5	12.2	14.6	14.8	14.3	13.3	12.6	13.7

Determine a 95% confidence interval for the waiting time.

Solution

The Excel function =T.INV(0.975; 9) obtains $t_{0.975;9} = 2.2621$. The average of the ten runs is $\bar{x} = 13.56$, and the deviation $s = 1.0616$. For $n = 10$; $\sqrt{10} = 3.1623$ then:

$$\bar{x} \pm t_{1-\frac{\alpha}{2}; n-1} * \left(\frac{s}{\sqrt{n}} \right)$$

$$13.56 \pm 2.2621 * (1.0616/3.1623) = 13.56 \pm 0.759$$

The average waiting time is estimated to be between 12.8 and 14.3 minutes.

Excel directly provides the function to determine the factor to construct the confidence interval:

=**CONFIDENCE.T** (Alpha; Deviation; Size).

=**CONFIDENCE.T**(0.05; 1.0616; 10) = 0.759

Goodness-of-Fit Tests

6

6.1 PROBABILITY MODEL FITTING

Studies involving statistical analysis require a correct evaluation of the assumptions of the procedures to be used. Most parametric statistical methods, such as analysis of variance, regression, quality control, reliability, t-test, estimation of confidence intervals for the population mean, among others, assume that the input data follow a normal distribution, which should be tested prior to the application of any statistical protocol. In simulation, the effect of using erroneous distributions is devastating to the credibility of the results. Consequently, it is necessary to verify compliance with the assumptions of the method under study.

When the entities of an unknown population exhibit random behavior, it can be determined whether their values are similar to those generated by a theoretical formal probability distribution function. Therefore, it is required to evaluate whether a given probability distribution can be used to represent the erratic values of an observed sample. The purpose of fitting a theoretical distribution to a set of unpredictable values is to identify a probability model that represents the observed values. In the extreme case, if a formal distribution of adequate fit cannot be found, the data can be summarized in an empirical distribution defined by probability weights.

6.2 MANUAL PROCEDURE

After obtaining a random sample of **n** values drawn from a population of interest to appreciate its shape, the data are sorted in ascending order and their sample statistics are estimated. In order to construct a frequency histogram the values are categorized or classified into mutually exclusive classes, with **k** class intervals or groups. A histogram plots the observed frequencies of the values of the variable. The shape of the histogram may suggest one or more classical probability distributions; or if no satisfactory probability model is found, the observed values can be characterized by an empirical distribution representing the values of the variable and their frequency ratios.

DOI: 10.1201/9781032701554-6

Generally, one or more formal probability distributions are found which can be adequately adjusted. Sometimes data are not available for a certain process, either because the system does not exist or because it is impossible to obtain observations. In these cases, the triangular distribution can be used as an approximation, assigning possible values to its three parameters: minimum, maximum, and most probable values. For example, the time required to carry out an activity can be approximated by consulting expert opinion, who can estimate the trio of parameters attributable to a triangular distribution.

Once one or more candidate distributions have been selected, a goodness-of-fit test allows the null hypothesis Ho to be formulated about the distribution model that best fits the data under scrutiny. A hypothesis is a presumptive statement which can be rejected or not. The null hypothesis defines the assumption intended to be rejected. Mutually exclusive statistical hypotheses are contrasted or tested against sample data.

The confidence level of the test $1 - \alpha$ measures the possibility of not rejecting an assertion that is true. The significance level α quantifies the type I error committed by rejecting an assumption that is true. The type II error β measures the possibility of no rejecting an assumption that is false. In everyday life not ruling out a false assumption and assuming it to be true is called a false positive. The power of the test $1 - \beta$ refers to the ability to detect as false and reject a hypothesis that is false (Table 6.1).

TABLE 6.1   Probability of choosing H_1 given that H_0 is true = α

	ASSERTION (H_0) IS TRUE	ASSERTION (H_0) IS FALSE
No Reject H_0 (H_0 is Chosen)	$1 - \alpha$ (confidence level)	β type II error (false negative)
Reject H_0 (H_1 is Chosen)	α type I error (significance level) (false positive)	$1 - \beta$ test power

6.3 GOODNESS-OF-FIT TESTS

Goodness-of-fit procedures are fundamental to determine the behavioral patterns of the various input processes to the model, as well as the validation of experimental results compared to those collected from the real system. The goodness-of-fit test determines whether statistically a series of values can be assumed to come from an independent sample generated from a theoretical probability distribution.

The test evaluates how good a fit is achieved between the frequency of occurrence of observations from an observed sample and the expected frequencies that would be obtained from the hypothesized distribution. This procedure compares two ordered sets of values arranged in frequencies: the observed sample values and the expected values generated by the candidate probability distribution. If the difference between the frequency of the observed and the expected is small, then the observed and expected values could be obtained from the hypothetical distribution.

In summary, the results of a goodness-of-fit test judge the approximation of the formal probabilistic model to the real observed situation. However, the rejection or not of a given distribution is subjective, since different probability distribution models may fit the same data set. The goodness-of-fit test is a useful guide to evaluate the suitability of a distribution. According to Banks (1998): "With few data, the goodness-of-fit test tends not to reject any candidate distribution. With many data, the test tends to reject the distribution of interest".

In addition, the ability to detect differences between the observed and the theoretical depends on several factors: number of classes **k** grouping the data, width of the class intervals, number of sample values, and shape of the candidate distribution.

The commercial software market offers several programs specialized in assisting the statistical evaluation of the best fit. These programs are generally based on the Chi-square, Kolmogorov-Smirnov, and Anderson-Darling methods. Most general-purpose statistical packages have the option of performing goodness-of-fit tests automatically. Other packages specialized in fit, such as Statfit, Expert Fit, @RISK, EasyFit, Simple Analyzer, Arena's Input Analyzer, among others, also allow to automatically fit theoretical distributions to the sample data, in addition to ranking the models that best represent them. These programs differ in the methods and criteria used to evaluate the best fit.

The quality of the fit depends significantly on the method used. To estimate the parameters of the candidate distribution, the best known methods are as follows: (a) maximum likelihood, (b) method of moments, and (c) least squares. However, the method of moments minimizes the mean squared error. In addition, each statistical package provider has its own secret calculation algorithms. Therefore, for the same set of data, a given software may evaluate the distributions it proposes in a certain order, while another package for the same data generally offers a different selection and ranking of the distributions it recommends.

The user can evaluate different optional settings. Since the selection criterion is a personal one, it is recommended to evaluate the "smoothness" of the fit. Moreover, if the difference between one distribution and another, in terms of for example, the mean square error is only a few thousandths, the recommendation is to select among the best candidates, the distribution that is most familiar to the user. Two traditional tests used to assess goodness-of-fit are presented below for illustrative purposes.

6.4 STATISTICAL TESTS ON THE FIT HYPOTHESES

After a descriptive evaluation of the data, a goodness-of-fit test is performed. The null hypothesis, **Ho**, states the suspicion about the possible distribution that could represent the data. Initially, a null hypothesis and an alternative hypothesis are proposed:

H_0: The observed values correspond to those generated from a certain distribution.
H_1: The observed values do not correspond to those generated from a certain distribution.

6.4.1 Chi-Square Test

This test is applicable to both discrete and continuous variable values. The n available values are grouped into k frequency intervals called classes. Defining the number of classes and the amplitude of each class affects the sensitivity of the test. This is evidence of a certain subjectivity of the test. However, it is considered a robust test if it is defined for **n** ≥ 30 values, more than two categories or classes and at least five observations per class. Cochkran (1952) recommends between 10 and 25 classes of equal length.

The Chi-square test statistic is obtained according to the following relationship:

$$\chi_o^2 = \sum_k \frac{(F_E - F_O)^2}{F_E}$$

where

F_O = observed frequency for each class interval or different value in the sample;
F_E = expected frequency of values in each class interval of the proposed distribution;
k = number of class intervals, each interval must contain at least five values.

6.4.2 Determination of the Number of Classes

To determine the number **k** of classes, of **n** values, it is customary to use the expressions: $k = \sqrt{n}$ or the Sturges relation: $k = 1 + 3.32 * \log(n;10)$.

Shorack (1963) offers a well-founded method for determining **k**: twice the fifth root of the square of the number of observations: $2n^{2/5}$.

Another option for constructing histograms is Scott's (1979) expression. Scott's relation calculates the optimal width of the class interval, by means of the expression:

$$A = 3.49 * S * n^{-1/3}$$

where **n** is the number of values, and **S** is the standard deviation of the **n** values. By dividing the range of the data by **A**, the number of classes **k** is obtained.

6.4.3 Rejection of the Fit

When the value of the Chi-square expression yields a large value, the test tends to reject the **Ho** hypothesis. Therefore, a right-tailed test is considered.

If the observations follow the distribution proposed in H_0, then it has approximately a Chi-square distribution, with $v = k - 1 - p$ degrees of freedom, where **p** corresponds to the number of parameters of the theoretical distribution to be estimated using the observed data. For example, if the fit is to a normal distribution, it is required to use the data to estimate its mean and variance, two parameters, then **p** = 2. If the distribution to fit is exponential only one parameter will be estimated from the data, then **p** = 1.

If $\chi_O^2 = 0$, the observed frequencies F_O match the frequencies of the theoretical distribution F_E and the fit is 100% perfect. Too much an ideal. If $\chi_O^2 > 0$, it should be compared against the theoretical value χ_E^2 to find out if the difference between the sample values and those from the distribution can be attributed to chance or random factors. The value is obtained for a significance level α and the corresponding degrees of freedom, where k corresponds to the number of class intervals and p to the number of parameters of the distribution which must be estimated from the observed values.

The value α is the probability of rejecting the null hypothesis when it is true; this is the type I error. It is usually stated as $\alpha = 0.05$. If the p-value of the test is less than α, the H_0 is rejected. That is, you have a confidence level of 95% of making the correct decision.

In Excel to obtain the theoretical value χ_E^2 we use the function:

=**CHISQ.INV.RT**(probability, deg_freedom **v**)

which corresponds to the inverse of the right tail probability of the Chi-square distribution. For $\alpha = 0.05$ and v = 7, $\chi_E^2 = 14.067$.

6.4.4 Kolmogorov-Smirnov (K-S) test

This test assumes that the distribution to be fitted is continuous. According to Law and Kelton (1991), it should only be applicable to continuous distributions, and when all the parameters of the distribution are known. This condition restricts its use. Both authors recommend its application with parameters estimated from the data, only in Exponential, Normal, Lognormal, Weibull, and Loglogistic distributions. Nevertheless, the K-S test has been used for discrete and continuous distributions with parameters obtained from the data. In general, Stephens (1974) argues that K-S is more powerful than the χ^2 test.

The K-S method is based on the interval in which the theoretical and observed cumulative distributions have the largest deviation or absolute difference. That is, it finds the maximum distance between the data sample distribution and the theoretical probability distribution. The maximum deviation is compared against the tabulated critical value of K-S to determine whether this difference can be attributed to random factors. The sample size corresponds to the number of degrees of freedom n.

Having defined the hypothesis H_o: The observed values X_i follow a given probability distribution, the K-S test is verified by means of the contrast statistic calculated according to the following relation:

$$D = Sup|F_E(x_i) - F_O(x_i)|, \quad para \ 1 \le i \le n$$

which obtains the maximum absolute value of the difference between the cumulative frequency of the expected or theoretical distribution $F_E(x_i)$, assumed in hypothesis H_o, and the cumulative frequency of the observed data $F_O(x_i)$.

The cumulative distribution is the probability of obtaining values less than or equal to a reference value X. The decision criterion is: If $D \le D_\alpha$ it is not rejected because we can to assume the chosen distribution represents the xi values. If **Ho** $D > D_\alpha$ it is rejected with significance level α. That is, if **Ho** is true then the probability of rejecting it is α.

After adjustment, commercials statistical packages automatically provide the **p**-value coefficient. If the **p**-value is less than α, **Ho** is rejected. On the other hand, if the **p**-value is greater than or equal to the significance level α, we conclude not to reject **Ho**.

6.5 QUANTILE-QUANTILE PLOTS

The quantile-quantile (Q-Q) or accumulated probability P-P plots can be used to evaluate whether two sets of values follow the same probability distribution.

A Q-Q plot presents the cumulative quantiles of the candidate distribution **Y** against the cumulative quantiles of the observed values **X**. If most of the points on the graph lie on a straight line **Y = X**, the data have the same distribution. The quantiles **Q** position the values of the distribution in parts, so that a proportion **p** of values is less than or equal to the quantile value. That is, until the value of the variable accumulates a certain proportion of values.

Thus, the 25th quantile corresponds to the value of the variable in which 25% of the values of the population are accumulated; and the median or 50 quantile is the value that positions the accumulated 50% of the values.

The Q-Q plot compares each observed quantile with the quantile of the theoretical distribution. If the plotted values follow the same distribution, the Q-Q plot is linear. To make a Q-Q plot the set of observed values is sorted in ascending order. Then, sample quantiles are calculated and compared to the theoretical quantile.

Many statistical tools assume normality of the data to be evaluated. To determine whether the observed values satisfy the condition of being normally distributed, the MATLAB software offers the function **qqnorm**(x). This function plots the quantiles of the observed values, sorted in ascending order and contained in the vector **x**, against the theoretical quantiles of the normal probability distribution. If the normal distribution approximates the data contained in **x** the plot approximates a straight line.

Example 6.1: Drawing a Q-Q Chart
Design a Q-Q plot to evaluate whether the following 50 data (Table 6.2) can be fitted to a normal distribution of mean 8.4 Liters/day and variance 2.12.

Solution

- Start by sorting the series of values from smallest to largest. From the ribbon: **Data / Sort /** select the range and **OK**
- Assign a position index **i** to each data. In this case from **i** = 1 to **i** = 50.
- Calculate the observed quantiles of the sample by means of the relation

$$\frac{\mathbf{i} - 0.3}{\mathbf{n} + 0.4}$$

TABLE 6.2   Liters/day to fit a normal distribution

4.3	6.7	7.7	9.1	10.6
4.5	6.8	7.9	9.2	10.7
4.8	7.0	8.0	9.5	10.8
5.2	7.0	8.2	9.9	11.3
5.2	7.1	8.3	10.0	11.4
6.1	7.2	8.5	10.0	11.5
6.2	7.2	8.5	10.1	11.5
6.2	7.4	8.5	10.1	11.6
6.4	7.4	8.8	10.2	11.9
6.4	7.5	8.9	10.4	12.4

Obtain the values of a normal distribution for the calculated **Q** quantiles of mean and deviation estimated from the sample of 50 values. For observed quantiles **X, i** = 1 to **i** = 50, calculate $(i - 0.3)/(50 + 0.4)$. For expected quantiles **Y** obtain average of 50 values: $\mu = 8.44$ and deviation $\sigma = 2.12$.

Figure 6.1 shows the layout of the Excel formulas suitable for obtaining the indices, the observed quantiles and the expected normals.

You want Excel to insert a scatter plot **XY** with the observed data column and the values column. Recall that Excel takes the values from the leftmost column for the Y-axis and the values from the other column for its X-axis.

	A	B	C	D
1	Index i	Liters/day	Quantiles (Observed) X	Q Normal (Expected) Y
2	1	4.3	=(A2-0.3)/(50.4)	=NORM.DIST(B2;8.44;2.12;1)
3	2	4.5	=(A3-0.3)/(50.4)	=NORM.DIST(B3;8.44;2.12;1)
4	3	4.8	=(A4-0.3)/(50.4)	=NORM.DIST(B4;8.44;2.12;1)
5	4	5.2	=(A5-0.3)/(50.4)	=NORM.DIST(B5;8.44;2.12;1)
6	5	5.2	=(A6-0.3)/(50.4)	=NORM.DIST(B6;8.44;2.12;1)
7	6	6.1	=(A7-0.3)/(50.4)	=NORM.DIST(B7;8.44;2.12;1)
49	48	11.6	=(A49-0.3)/(50.4)	=NORM.DIST(B49;8.44;2.12;1)
50	49	11.9	=(A50-0.3)/(50.4)	=NORM.DIST(B50;8.44;2.12;1)
51	50	12.4	=(A51-0.3)/(50.4)	=NORM.DIST(B51;8.44;2.12;1)

FIGURE 6.1   Formulas for indices, quantiles, and expected normal.

To obtain the graph, the following procedure is performed in Excel (Figure 6.2):

Select data range / **INSERT/** on Charts **Insert Scatter (X, Y)**

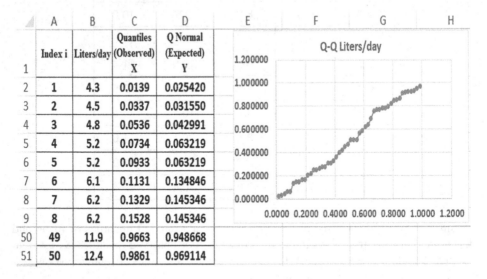

	A	B	C	D	E	F	G	H
	Index i	Liters/day	Quantiles (Observed) X	Q Normal (Expected) Y				
1								
2	1	4.3	0.0139	0.025420				
3	2	4.5	0.0337	0.031550				
4	3	4.8	0.0536	0.042991				
5	4	5.2	0.0734	0.063219				
6	5	5.2	0.0933	0.063219				
7	6	6.1	0.1131	0.134846				
8	7	6.2	0.1329	0.145346				
9	8	6.2	0.1528	0.145346				
50	49	11.9	0.9663	0.948668				
51	50	12.4	0.9861	0.969114				

FIGURE 6.2 Q-Q chart.

6.6 PROB-PROB CHARTS

A probability-probability (P-P) chart plots the cumulative probability function of the observed values against the candidate theoretical cumulative probability distribution. To draw it, the observed values are ordered in ascending order. A column containing the position or **i**th order indices from **1** to **n** is arranged. Another column will contain the column **i/n**, value to be placed on the **X**-axis, where **n** is the number of values and **i** is the i-th order. On the **Y**-axis, the cumulative probabilities of the candidate or theoretical cumulative distribution function **F**(x) are arranged. If the plot corresponds to a straight line, then the observed values can be represented by the assumed theoretical distribution **f**(x) (Figure 6.3).

Example 6.2: P-P Chart for a Normal Distribution
The P-P plot shows the series of points on a straight line, from which it is concluded that the values of **X** fit adequately to a normal probability distribution of parameters 98.79 for the mean and 19.21 for the standard deviation.

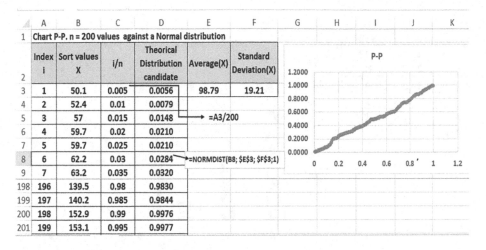

	A	B	C	D	E	F	G	H	I	J	K
1	Chart P-P. n = 200 values against a Normal distribution										
2	Index i	Sort values X	i/n	Theorical Distribution candidate	Average(X)	Standard Deviation(X)					
3	1	50.1	0.005	0.0056	98.79	19.21					
4	2	52.4	0.01	0.0079							
5	3	57	0.015	0.0148	=A3/200						
6	4	59.7	0.02	0.0210							
7	5	59.7	0.025	0.0210							
8	6	62.2	0.03	0.0284	=NORMDIST(B8; $E$3; $F$3;1)						
9	7	63.2	0.035	0.0320							
198	196	139.5	0.98	0.9830							
199	197	140.2	0.985	0.9844							
200	198	152.9	0.99	0.9976							
201	199	153.1	0.995	0.9977							

FIGURE 6.3 P-P chart.

6.7 CHI-SQUARE FIT: POISSON DISTRIBUTION

Example 6.3: During a span of 509 weeks, the number of accidents that occurred at a road intersection was recorded. The 509 weekly records were grouped according to the number of accidents/week as shown in Table 6.3.

To fit a probability distribution, we start with a histogram that allows us to evaluate graphically the possible shape of the available data (Figure 6.4).

Since this is a discrete variable, Poisson behavior is suspected. Once one or more theoretical distributions that are considered to fit the data have been identified, the parameters of the candidate(s) will be determined.

Usually, the variance and sample mean are calculated with the following formulas:

$$S^2 = \frac{\sum_{i=1}^{i=n} X_i^2 F_i - n\bar{X}^2}{n-1} = \frac{440 - (509 * 0.5147^2)}{508} = 0.6007$$

TABLE 6.3 Accidents at a road intersection.

X ACCIDENTS PER WEEK	NUMBER OF WEEKS WITH X ACCIDENTS (FREQUENCY)
0	315
1	142
2	40
3	9
4	2
5	1

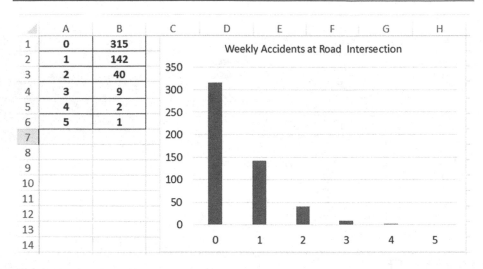

FIGURE 6.4   Histogram for the data in Table 6.2.

$$\bar{X} = \frac{\sum_{i=1}^{i=n} X_i F_i}{n} = \frac{262}{509} = 0.5147$$

In addition to the graph, the values of the mean and variance are observed. The similarity of both statistics increases the expectation that the best fitting distribution is a Poisson. In theory, the Poisson has identical mean and variance.

To determine the value of the average Poisson parameter, both values are averaged, $(0.5147 + 0.6007)/2 = 0.5577$. From the observed data, we estimate the average Poisson as 0.577. Consequently:

Null Hypothesis Ho: The number of accidents occurring weekly at the intersection can be represented by a Poisson distribution with average 0.5577 accidents/week.

Alternate Hypothesis H₁: The number of accidents occurring weekly at the intersection cannot be represented by a Poisson distribution with mean 0.5577.

It is observed that the accident event can occur equally in any interval. Furthermore, the occurrence of one event does not exert any influence or effect on whether or not another event occurs. Moreover, the probability of zero occurrences in any interval is high; the average number of occurrences in an interval is small; all this and the shape of the relative frequency plot reinforces the idea of fitting a Poisson model.

By choosing a Poisson of average 0.5577, the model corresponding to the weekly number of accidents is (Table 6.4):

$$P(x) = \frac{\lambda^x * e^{-\lambda}}{x!} = \frac{0.5577^x * e^{-0.5577}}{x!}$$

TABLE 6.4   Application of χ^2 fit: number of accidents/week

X	POISSON THEORETICAL PROBABILITY P_n	EXPECTED FREQUENCY F_E = 509 * P_n	OBSERVED FREQUENCY F_O	$\frac{(F_O - F_E)^2}{F_E}$
0	0.571	291	315	1.98
1	0.319	162	142	2.47
2	0.089	45	40	0.56
3	0.017	9	9	0
4	0.003	1	2	0.09
5	0.001	1	1	0
Total		509	509	6.10

Since some classes have less than 5 values, classes 4 and 5 are grouped in class 3.

TABLE 6.5   Grouping of classes 4 and 5 in class 3

X	POISSON THEORETICAL PROBABILITY P_n	EXPECTED FREQUENCY F_E = 509 * P_n	OBSERVED FREQUENCY F_O	$\frac{(F_O - F_E)^2}{F_E}$
0	0.571	290.64	315	2.041
1	0.319	162.52	142	2.59
2	0.089	45.32	40	0.624
3	0.019	9.73	12	0.53
		Total:	509	$\chi^2 = 5.78$

The Excel **CHISQ.INV.RT(α;v)** function with α = 0.05; and with v = 2 degrees of freedom, (v = 4 – 1 – 1), returns the inverse of the probability α, right tailed of the Chi-square distribution. In this case, the critical tabular value is:

$$= \textbf{CHISQ. INV. RT}(0.05; 2) = \chi_E^2 = 5.99$$

Since 5.78 < 5.99, the calculated value is less than the tabulated critical value, **H₀** is not rejected. That is, there is no significant difference between the observed frequency distribution and those from a Poisson distribution with mean 0.5577.

Therefore, with 95% confidence, the number of accidents at the intersection can be modeled by a Poisson distribution of average 0.5577 accidents per week.

6.8 KOLMOGOROV-SMIRNOV FIT: POISSON DISTRIBUTION

Example 6.4: Using the data in Table 6.5 perform the Kolmogorov-Smirnov test for a Poisson distribution.

Solution

Although the assumptions of the K-S test include randomness of the sample and continuity of the candidate theoretical distribution; for practical purposes only, disregarding the non-continuity of the number of accidents and the warning of Law and Kelton, on the intersection accident data the K-S test will be used. Although Gottfried (1963) shows that the **D** statistic in the K-S test produces conservative results when applied to discrete distributions, this does not limit the use of K-S to the continuous case only.

Table 6.6 shows that the largest absolute difference is 0.048 with zero accidents. Therefore, using formula 6.2 one has that D = 0.048. To find the contrast value for **n** = 509 and α = 0.05 from the K-S table one obtains:

TABLE 6.6   K-S fit for the number of accidents

NUMBER OF EVENTS X	OBSERVED FREQUENCY F_O	OBSERVED PROBABILITY $P(F_O) = F_O/509$	OBSERVED CUMULATIVE PROBABILITY (1)	THEORETICAL PROBABILITY $F(X)$	THEORETICAL CUMULATIVE PROBABILITY $F(X)$ (2)	DIFFERENCE (1) – (2)
0	315	0.619	0.619	0.571	0.571	0.048
1	142	0.279	0.898	0.319	0.890	0.008
2	40	0.078	0.976	0.089	0.979	0.003
3	9	0.018	0.994	0.017	0.996	0.002
4	2	0.004	0.998	0.003	0.999	0.001
5	1	0.002	1.000	0.001	1.000	0.000

$$D_T = \frac{1.36}{\sqrt{509}} = 0.0603$$

For values of **n** greater than 50, the values of D_T from the K-S table are obtained from the recurrence ratio D(0.05; 509) = 0.0603 at 5% significance level.

Since the largest deviation was 0.048, and this is less than 0.063, the conclusion is: do not reject **Ho**. It is concluded that the weekly number of accidents at the intersection can be represented by a Poisson distribution with an average of 0.5577 accidents per week.

6.9 χ^2 Fit: Uniform Distribution

Example 6.5: A lottery game consists of matching three sequences of three numbers selected from a wheel with values between 0 and 9. From the record of several draws of lottery game, 500 digits corresponding to the sequences of winning numbers were obtained. Table 6.7 shows the observed frequency of each digit.

TABLE 6.7   Observed frequencies in Triple Jackpot

DIGIT	0	1	2	3	4	5	6	7	8	9	TOTAL
Observed Frequency	62	58	36	28	40	70	60	40	72	34	500

Hypothesis:

Ho: The digits of the lottery draws do not present significant difference with those generated by a uniform distribution for a confidence level of 99%.

H$_1$: The digits of the lottery draws present a significant difference with those generated by a uniform distribution for a confidence level of 99%.

Since the digits from 0 to 9 should have the same probability of occurring, the probability of occurrence of any of them is 0.10; so the expected number of times that any value should occur is 0.10 multiplied by 500, which gives 50. That is, in 500 individual numbers it is expected to get 50 zeros, 50 ones, 50 twos, etc. (Table 6.8).

TABLE 6.8   Expected frequency per digit

DIGIT	0	1	2	3	4	5	6	7	8	9	TOTAL
Expected Frequency	50	50	50	50	50	50	50	50	50	50	500

The chi squared test is applied to evaluate the fit to a uniform distribution. Table 6.9 shows the calculations for the test. Row 4 presents the value of the test statistic $\chi^2 = 46.6$. Then for $v = 10 - 1 = 9$ degrees of freedom and $\alpha = 0.01$, the value of χ_E^2 tabulated is **CHISQ.INV.RT(0.01;9) = 21.666**. Since $\chi_O^2 > \chi_E^2$, hence **Ho** is rejected.

TABLE 6.9   Adjustment χ^2 in Triple Jackpot Lotto

DIGIT	0	1	2	3	4	5	6	7	8	9	TOTAL
Observed Frequency F$_o$	62	58	36	28	40	70	60	40	72	34	500
Expected Frequency F$_E$	50	50	50	50	50	50	50	50	50	50	500
$\frac{(F_E - F_o)}{F_E}$	2.88	1.28	3.92	9.68	2	8	2	2	9.68	6.12	46.6

It is concluded that the process is biased. The digits recorded belonging to the observed draws of the Triple Jackpot present significant difference with those generated by a uniform distribution for a confidence level of 99%.

6.10 FITTING χ^2: BINOMIAL DISTRIBUTION

Example 6.6: Six coins were tossed in the air 1,000 times. The number of faces obtained in each toss was recorded. The number of tosses in which 0, 1, 2, 3, 3, 4, 5, and 6 faces were obtained is shown in Table 6.10. Determine if the number of faces obtained in each throw can be represented by a Binomial distribution at the 95% confidence level.

TABLE 6.10   Faces in 1,000 tosses of six coins

FACES	0	1	2	3	4	5	6
Tosses	23	101	231	305	225	95	20

Solution
It is necessary to estimate from the observed data the probability **p** of obtaining heads. The number of repetitions is **n** = 6 tosses. The average number of heads is estimated from the observed frequency distribution:

$$\bar{X} = \frac{\Sigma f_x * X}{\Sigma f} = \frac{23 * 0 + 101 * 1 + 231 * 2 + 305 * 3 + 225 * 4 + 95 * 5 + 20 * 6}{1,000} = 2.973$$

In a Binomial distribution the average is: $\mu = E(x) = n * p = 6 * p$, from where:

$$p = \frac{\mu}{6} = \frac{2.973}{6} = 0.4955$$

To obtain the expected values the Binomial distribution formula is used:

$$B(x, p, n) = \frac{n!}{x!(n-x)!} p^x (1-p)^{n-x}$$

With the parameter estimated to the data, the procedure of evaluation or assessment of the hypothesis of binomiality of the data is shown below.

Alternative hypothesis H_1: The number of faces does not fit a Binomial distribution of parameter **p** = 0.4955.
Null hypothesis Ho: The number of heads obtained when tossing 6 coins can be represented by a Binomial distribution of parameter **p** = 0.4955.

The Binomial distribution function, where **x** represents the number of heads obtained when tossing six coins simultaneously, each coin with approximate probability to obtain a head of 0.5, would be:

$$B(x; 0.4955; 6) = \frac{6!}{x!(6-x)!} * (0.4955)^x * (0.5045)^{6-x}$$

Using the above formula, we can estimate the probability of obtaining **x** heads, depending on the value of the random variable. Although it is better to use the binomial statistical function of Excel:

= BINOM.DIST (x; n; p; cumulative)

= BINOM.DIST(0; 6; 0.4955; 0) = 0.0165 corresponding to the probability of not getting heads when tossing 6 coins.

The probability of getting 0 faces multiplied by 1,000 tosses gives the expected or theoretical value. For example, for **x** = 0, the expected frequency of not getting heads is 0.016∗1,000 = 16.48 tosses. The results are summarized in Table 6.11.

TABLE 6.11 Expected and observed frequency of tossing six coins

NUMBER OF FACES (X)	P (X FACES)	EXPECTED FREQUENCY (FE)	OBSERVED FREQUENCY (F$_o$)
0	0.01648	16.5	23
1	0.09716	97.2	101
2	0.2386	238.6	231
3	0.3124	312.4	305
4	0.2301	230.1	225
5	0.0904	90.4	95
6	0.0148	14.8	20

To determine the degrees of freedom, we have seven classes. One will be subtracted because the data was used to estimate the proportion of the population: the parameter **p**. Then, the degrees of freedom are obtained by: v= Classes–1 – Estimated parameters = 7–1–1 = 5.

For α = 0.05 and 5 degrees of freedom, the value of χ_E^2 = **CHISQ.INV.RT**(0.05; 5) = 11.07. Therefore, if the calculated value χ_o^2 < 11.07 **Ho** is not rejected. If χ_o^2 >11.07 **Ho** is rejected. The test statistic is calculated from:

$$\chi_o^2 = \sum_\forall \frac{(Fe - Fo)^2}{Fe}$$

$$\frac{(16.5 - 23)^2}{16.5} + \frac{(97.2 - 101)^2}{97.2} + \frac{(238.6 - 231)^2}{238.6} + \frac{(312.4 - 305)^2}{312.4}$$

$$+ \frac{(230.1 - 225)^2}{230.1} + \frac{(90.4 - 95)^2}{90.4} + \frac{(14.8 - 20)^2}{14.8} = 5.3$$

Since $\chi_0^2 = 5.3$ is less than $\chi_E^2 = 11.07$ then **Ho** is not rejected. It is concluded with 95% confidence, that the number of heads when tossing 6 balanced coins can be modeled with a Binomial (x; 0.4955; 6) distribution.

6.11 χ^2 FIT: NORMAL DISTRIBUTION

Example 6.7: Determine if the recorded weight of 90 bulls can be approximated by a normal distribution with mean $\mu = 501.37$ kilograms, standard deviation $\sigma = 14.39$ and 95% confidence level.

500.5	495.2	506.3	502.7	502.5	503.6	480.2	511.1	519.5
508.2	491.7	504.8	502.8	495.1	510.9	529.9	489.7	501.1
519.1	507.6	500.8	516.2	508.2	473.9	506.6	521.1	494.2
480.3	510.3	505.9	519.2	517.5	490.5	488.9	502.7	518.6
488.8	500.8	522.3	485.9	507.1	495.2	468.3	488.3	488.2
498.7	503.4	467.2	491.7	524	506.7	483.9	514.4	481.9
507.2	512.1	501	505	494	494.3	478.2	493	500.7
517.5	465.4	525.5	486.3	477.4	518.6	508.9	495.4	485.4
501.9	502.4	503.4	519	497.8	519.7	508.2	481.3	501.8
499.5	505.5	503.5	500	496.9	529.1	529.8	517.3	485.7

From the Excel sheet, Figure 6.5, cell E3, we obtain the range of the data, 64.5 kilograms, and estimate the number of class intervals corresponding to the root of 90 (9.5). From the data, the two parameters of the distribution to be evaluated $\mu = \bar{X}$ and $\sigma = s$ are estimated, which will reduce the degrees of freedom of the Chi-square distribution by 2.

Since, in a normal distribution the mode, mean, and median of the data theoretically are equals, it is convenient to estimate them from data and verify if there is tendency to coincide. Moreover, data histogram should show symmetrical trend; the elaboration of a whisker plot can show the normal trend. Also, 95% of the values should be within two deviations of the mean. In Figure 6.5, the histogram suggests a normal function.

The following hypotheses will be tested:

Ho: The weight of bulls from the batch of 90 animals, can be represented by a normal distribution of mean 501.37 kilograms and deviation 14.39.

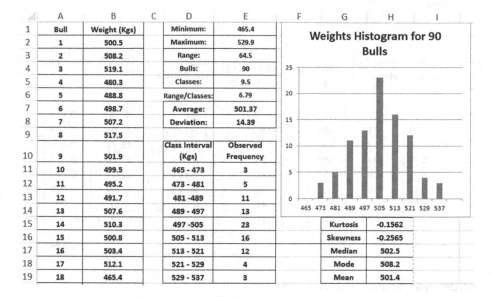

FIGURE 6.5 Weight of first 18 animals plus histogram.

H$_1$: The weight of bulls from the batch of 90 animals is not representable by a normal distribution of mean 501.37 kilograms and deviation 14.39.

To obtain the expected probabilities corresponding to the proposed probability distribution, the weights **x** are standardized by the following conversion:

$$z = \frac{x - \mu}{\sigma} = \frac{x - 501.37}{14.39}$$

The first value of **x** is 473 and the last 537. Figure 6.6 shows the calculations until the calculated Chi-square value is obtained. It can be seen that the frequencies observed in two class intervals, 465–473 and 529–537, have frequencies less than five values; this requires grouping some frequencies with the neighboring interval.

The Chi-square test statistic is obtained as follows:

$$\chi_o^2 = \sum_{\forall \; class \; Interval} \frac{(Fe - Fo)^2}{Fe} = 1.687$$

The number of degrees of freedom **v** is obtained with v = Classes − p − 1 = 7 − 2 − 1 = 4.

The value of χ_E^2 for 95% confidence and 4 degrees of freedom is achieved from Excel as:

=**CHISQ.INV.RT**(0.05;4) = 9.4877

F2	▼	:	×	✓	fx	=E2*90

	A	B	C	D	E	F	G
1	Class Interval (Kgs)	Observed Frequency Fo	Limits X	Standard Values Z (X-501.37)/14.39	Expected Probabilities P(Z)	Expected Frequency Fe 90*P(Z)	$\dfrac{(Fe - Fo)^2}{Fe}$
2	465-481	8	481	-1.42	0.078	7.06	0.125
3	481-489	11	489	-0.86	0.117	10.49	0.025
4	489-497	13	497	-0.30	0.186	16.71	0.824
5	497-505	23	505	0.25	0.219	19.70	0.553
6	505-513	16	513	0.81	0.191	17.18	0.082
7	513-521	12	521	1.36	0.123	11.09	0.075
8	521-537	7	537	2.48	0.080	7.17	0.004
9				(C2-501.37)/14.39		Sum X_0^2 =	1.687
10					=NORM.S.DIST(D2;1)		
11							=((F2-B2)^2)/F2
12				NORM.S.DIST(D3;1) - NORM.S.DIST(D2;1)			

FIGURE 6.6   Normal fit for the weight of 90 bulls.

Since the calculated Chi-square value of 1.687, Figure 6.6, cell G9, is less than the expected value of 9.48, the null hypothesis is not rejected. It is concluded that the recorded weight of the 90 bulls, with 95% confidence, can be fit to a normal distribution of mean 501.37 kilograms and variance of 14.39 with 95% confidence.

Figure 6.7 shows the formulas used to obtain the results of Figure 6.6.

	A	B	C	D	E	F	G
1	Class Interval (Kgs)	Observed Frequency Fo	Limits X	Standard Values Z (X-501.37)/14.39	Expected Probabilities P(Z)	Expected Frequency Fe 90*P(Z)	$\dfrac{(Fe - Fo)^2}{Fe}$
2	465-481	8	481	=(C2-501.37)/14.39	=NORM.S.DIST(D2;1)	=E2*90	=((F2-B2)^2)/F2
3	481-489	11	489	=(C3-501.37)/14.39	=NORM.S.DIST(D3;1) - NORM.S.DIST(D2;1)	=E3*90	=((F3-B3)^2)/F3
4	489-497	13	497	=(C4-501.37)/14.39	=NORM.S.DIST(D4;1) - NORM.S.DIST(D3;1)	=E4*90	=((F4-B4)^2)/F4
5	497-505	23	505	=(C5-501.37)/14.39	=NORM.S.DIST(D5;1) - NORM.S.DIST(D4;1)	=E5*90	=((F5-B5)^2)/F5
6	505-513	16	513	=(C6-501.37)/14.39	=NORM.S.DIST(D6;1) - NORM.S.DIST(D5;1)	=E6*90	=((F6-B6)^2)/F6
7	513-521	12	521	=(C7-501.37)/14.39	=NORM.S.DIST(D7;1) - NORM.S.DIST(D6;1)	=E7*90	=((F7-B7)^2)/F7
8	521-537	7	537	=(C8-501.37)/14.39	=NORM.S.DIST(D8;1) - NORM.S.DIST(D7;1)	=E8*90	=((F8-B8)^2)/F8
9						Sum X_0^2 =	=SUM(G2:G8)

FIGURE 6.7   Formulas for the normal fit to the weight of 90 bulls.

6.12 χ^2 FIT: EXPONENTIAL DISTRIBUTION

Example 6.8: In a simulation study the inter-arrival times of vehicles at an auto bank were recorded. Over a 90-minute time span 221 vehicles arrived. There are 220 inter-arrival times available. Table 6.12 contains the values obtained although they were previously sorted. Evaluate an adequate exponential fit with 95% confidence.

TABLE 6.12 Inter-arrival times of 221 cars

0.01	0.01	0.01	0.01	0.01	0.01	0.01	0.01	0.02	0.02	0.03	0.03	0.03	0.04
0.04	0.04	0.04	0.04	0.04	0.05	0.05	0.05	0.05	0.05	0.05	0.05	0.05	0.05
0.05	0.06	0.06	0.06	0.06	0.07	0.07	0.07	0.07	0.07	0.07	0.07	0.07	0.07
0.07	0.08	0.08	0.08	0.08	0.09	0.09	0.10	0.10	0.10	0.10	0.10	0.10	0.10
0.10	0.10	0.11	0.11	0.11	0.11	0.11	0.12	0.12	0.12	0.12	0.13	0.13	0.14
0.14	0.14	0.14	0.15	0.15	0.15	0.15	0.15	0.15	0.17	0.18	0.19	0.19	0.19
0.20	0.21	0.21	0.21	0.21	0.21	0.22	0.22	0.22	0.23	0.23	0.23	0.23	0.23
0.24	0.25	0.25	0.25	0.25	0.25	0.26	0.26	0.26	0.26	0.26	0.27	0.28	0.28
0.29	0.29	0.30	0.31	0.31	0.32	0.35	0.35	0.35	0.36	0.36	0.36	0.37	0.37
0.38	0.38	0.38	0.38	0.38	0.39	0.40	0.40	0.41	0.41	0.43	0.43	0.43	0.44
0.45	0.45	0.46	0.47	0.47	0.47	0.48	0.49	0.49	0.49	0.49	0.50	0.50	0.50
0.50	0.51	0.51	0.51	0.52	0.52	0.53	0.53	0.53	0.54	0.54	0.55	0.55	0.56
0.57	0.57	0.60	0.61	0.61	0.63	0.63	0.64	0.65	0.65	0.65	0.69	0.69	0.70
0.72	0.72	0.72	0.74	0.75	0.76	0.77	0.79	0.84	0.86	0.87	0.88	0.88	0.90
0.93	0.93	0.95	0.97	1.03	1.05	1.05	1.06	1.09	1.10	1.11	1.12	1.17	1.18
1.24	1.24	1.28	1.33	1.38	1.44	1.51	1.72	1.83	1.96				

Solution

The average inter-arrival time is $\bar{x} = 0.399$ minutes and its variance $s^2 = 0.144$. An estimate of the coefficient of variation is:

$$cv = \frac{\sigma}{\mu} = \frac{\sqrt{0.144}}{0.399} = 0.951$$

This value is close to 1, suggesting an exponential fit. Furthermore, if histograms of the interarrival time are drawn starting from 0 and with increments of 0.05, 0.075, or 0.1, a graph with a tendency to exponential shape is obtained. See Figure 6.8.

After extracting the root of 221, the times are grouped into 15 class intervals. We will try to fit an exponential distribution whose density function is:

$$f(t) = \frac{1}{0.399} e^{\frac{-t}{0.399}}$$

And its cumulative probability function:

$$F(t) = 1 - e^{\frac{-t}{0.399}}$$

Ho: The times between the arrivals of vehicles at the auto bank can be represented, with a confidence level of 95%, by an exponential distribution with an average of 0.399 minutes.

	A	B	C	D	E	F	G	H	I	J	K
		Class Intervals	Observed Frequency		Expected Probability	Expected Frequency Fe	$(Fe - Fo)^2$			Arrivals Histogram	
2	0.01	(minutes)	Fo (cars)		P(t1< T <t2)	220*P(T)	Fe				
3	0.01	0.010	8		0.0248	5.4	1.199				
4	0.01	0.149	65		0.2869	63.1	0.056				
5	0.01	0.289	39		0.2037	44.8	0.755				
6	0.01	0.428	24		0.1426	31.4	1.729				
7	0.01	0.567	32		0.1006	22.1	4.393				
8	0.01	0.706	14		0.0710	15.6	0.169				
9	0.01	0.846	9		0.0504	11.1	0.396				
10	0.02	0.985	9		0.0353	7.8	0.196				
11	0.02	1.124	8		0.0249	5.5	1.157				
12	0.03	1.264	4		0.0177	3.9	0.003				
13	0.03	1.403	3		0.0124	2.7	0.028				
14	0.03	1.542	2		0.0087	1.9	0.003		$X_t^2 (0.05;13) =$ 22.4		
15	0.04	1.681	0		0.0062	1.4	1.357				
16	0.04	1.821	1		0.0044	1.0	0.001		Average Time = 0.3993		
17	0.04	and greater...	2		0.0044	1.0	1				
18	0.04					Sum $X_c^2 =$ 12.4					

FIGURE 6.8 Exponential fit test for inter-arrival time.

H_1: The times between vehicle arrivals at the self-bank cannot be represented, at the 95% confidence level, by an Exponential distribution with average 0.399 minutes.

Figure 6.8 shows the calculations for the fit and the histogram that represents them. An exponential behavior can be observed in the histogram. With $\alpha = 0.05$, av $= 15-1-1 = 13$ degrees of freedom the critical value is obtained: $\chi^2 = 22.362$.

Since the calculated χ_o^2 equals 12.4 and this is lower than the corresponding critical value X_T^2 (0.05;13) = **CHISQ.INV**(0.05;13) = 22.4, **Ho** is not rejected. Consequently, the inter-arrival times of vehicles at the auto bank can be represented, with 95% confidence level, by an exponential distribution of mean 0.399 minutes.

Figure 6.9 shows the formulas used to obtain the results of Figure 6.8.

	A	B	C	D	E	F	G
		Class Intervals	Observed Frequency			Expected Frequency	$\frac{(Fe-Fo)^2}{Fe}$
2	0.01	(minutes)	Fo (cars)		Expected Probability P(t1< T <t2)	Fe 220*P(T)	
3	0.01	0.01	8		=EXPONDIST(0.01;(1/0.399);1)	=220*E3	=((F3-C3)^2)/F3
4	0.01	0.149285714285714	65		=EXPONDIST(0.149;(1/0.399);1) - E3	=220*E4	=((F4-C4)^2)/F4
5	0.01	0.288571428571429	39		=EXPONDIST(0.289;(1/0.399);1) - EXPONDIST(0.149;(1/0.399);1)	=220*E5	=((F5-C5)^2)/F5
6	0.01	0.427857142857143	24		=EXPONDIST(0.428;(1/0.399);1) - EXPONDIST(0.289;(1/0.399);1)	=220*E6	=((F6-C6)^2)/F6
7	0.01	0.567142857142857	32		=EXPONDIST(0.567;(1/0.399);1) - EXPONDIST(0.428;(1/0.399);1)	=220*E7	=((F7-C7)^2)/F7
8	0.01	0.706428571428571	14		=EXPONDIST(0.706;(1/0.399);1) - EXPONDIST(0.567;(1/0.399);1)	=220*E8	=((F8-C8)^2)/F8
9	0.01	0.845714285714286	9		=EXPONDIST(0.846;(1/0.399);1) - EXPONDIST(0.706;(1/0.399);1)	=220*E9	=((F9-C9)^2)/F9
10	0.02	0.985	9		=EXPONDIST(0.985;(1/0.399);1) - EXPONDIST(0.846;(1/0.399);1)	=220*E10	=((F10-C10)^2)/F10
11	0.02	1.12428571428571	8		=EXPONDIST(1.124;(1/0.399);1) - EXPONDIST(0.985;(1/0.399);1)	=220*E11	=((F11-C11)^2)/F11
12	0.03	1.26357142857143	4		=EXPONDIST(1.264;(1/0.399);1) - EXPONDIST(1.124;(1/0.399);1)	=220*E12	=((F12-C12)^2)/F12
13	0.03	1.40285714285714	3		=EXPONDIST(1.403;(1/0.399);1) - EXPONDIST(1.264;(1/0.399);1)	=220*E13	=((F13-C13)^2)/F13
14	0.03	1.54214285714286	2		=EXPONDIST(1.542;(1/0.399);1) - EXPONDIST(1.403;(1/0.399);1)	=220*E14	=((F14-C14)^2)/F14
15	0.04	1.68142857142857	0		=EXPONDIST(1.681;(1/0.399);1) - EXPONDIST(1.542;(1/0.399);1)	=220*E15	=((F15-C15)^2)/F15
16	0.04	1.82071428571429	1		=EXPONDIST(1.821;(1/0.399);1) - EXPONDIST(1.681;(1/0.399);1)	=220*E16	=((F16-C16)^2)/F16
17	0.04	and greater...	2		=EXPONDIST(1.821;(1/0.399);1)- EXPONDIST(1.681;(1/0.399);1)	1	=((F17-C17)^2)/F17
18	0.04					Sum	=SUM(G3:G17)

FIGURE 6.9 Formulas used in Figure 6.8.

6.13 χ^2 FIT: WEIBULL DISTRIBUTION

Example 6.9: After expiration of the 1-year warranty for a special model, an automotive dealer recorded the time to first failure for 99 cars (Table 6.13).

Evaluate the fit of a Weibull distribution to represent the duration of car operation after the warranty lapse.

Solution
Developing a histogram, in addition to calculating descriptive statistics, helps to explore the fit of a Weibull distribution to the failure data of the 99 vehicles (Figures 6.10 and 6.11).

TABLE 6.13 Car running time (months)

11.4	15.5	5.0	16.6	17.4	10.5	19.4	9.5	3.7
21.8	23.9	22.6	19.7	16.5	19.3	11.9	7.8	25.5
8.0	24.1	27.4	20.8	17.5	30.7	6.1	3.5	12.8
37.6	42.8	10.3	28.9	34.3	3.9	19.8	24.2	25.0
11.5	20.1	9.8	15.0	11.8	12.4	15.1	10.2	6.4
6.7	5.0	3.0	25.3	17.5	2.2	14.1	6.8	13.1
26.5	12.3	10.1	11.9	22.9	14.5	18.7	14.6	5.2
7.9	8.9	18.9	50.4	7.0	25.1	18.5	17.8	34.5
36.9	20.9	11.4	6.7	21.1	28.1	5.6	21.6	26.0
26.2	36.2	18.6	21.4	3.3	3.7	20.0	19.2	8.4
9.8	27.4	35.1	8.8	4.0	8.0	33.8	9.4	34.7

	A	B	C	D	E	F	G	H	I	J	K	L	M	N
1	11.4	15.5	5	16.6	17.4	10.5	19.4	9.5	3.7		Min:	2.2		Bins
2	21.8	23.9	22.6	19.7	16.5	19.3	11.9	7.8	25.5		Max:	50.4		2.2
3	8	24.1	27.4	20.8	17.5	30.7	6.1	3.5	12.8		Range:	48.2		7.56
4	37.6	42.8	10.3	28.9	34.3	3.9	19.8	24.2	25		Classes	9		12.91
5	11.5	20.1	9.8	15	11.8	12.4	15.1	10.2	6.4		Wide:	5.36		18.27
6	6.7	5	3	25.3	17.5	2.2	14.1	6.8	13.1					23.62
7	26.5	12.3	10.1	11.9	22.9	14.5	18.7	14.6	5.2		=L2-L1			28.98
8	7.9	8.9	18.9	50.4	7	25.1	18.5	17.8	34.5			=L3/9		34.33
9	36.9	20.9	11.4	6.7	21.1	28.1	5.6	21.6	26					39.69
10	26.2	36.2	18.6	21.4	3.3	3.7	20	19.2	8.4		=N2+$L$5			45.04
11	9.8	27.4	35.1	8.8	4	8	33.8	9.4	34.7					50.40

FIGURE 6.10 Exploration of data and its indicators.

Once data times are on Excel, to build a histogram from the Excel ribbon:

DATA / Data Analysis / Histogram / OK

It is essential to estimate the scale and shape parameters of the Weibull distribution. Excel will be used and the least squares regression method will be applied, filling columns according to the following procedure (Table 6.14).

Figure 6.12 shows the calculations for the estimation of the parameters using Excel.

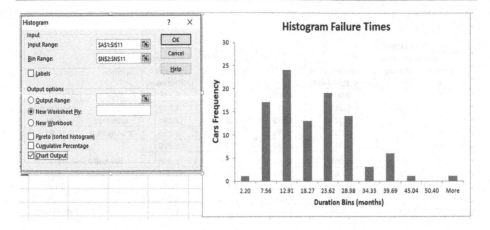

FIGURE 6.11 Histogram for operating times.

Once the parameters of scale 19.17 and shape 1.80 are obtained (Figure 6.12, cell H8 y H9), we proceed to obtain the calculated Chi-square value as shown in Figure 6.13.

Since frequency in each class should be at least five observations, it is required for chi-square procedure to group some class intervals as it is shown in Figure 6.14. Observed frequency values in cells D10, D11, and D12 (8+1+1) are grouped in cell D19. Also the values contained in cells D5 and D6 (1+28) are grouped in the value 29 (Figure 6.14, cell D15).

TABLE 6.14 Estimation of Weibull parameters

1. Sort the values of the variable in ascending order.
2. Assign an increasing index number i to each value of the variable i = 1, 2, 3, ...
3. Obtain the position or Median Ranks, according to:
   MR = $(i - 0.3)/(n + 0.4)$.
4. Calculate $1/(1 - MR)$.
5. Calculate the double logarithm $\ln(\ln(1/(1 - MR)))$.
6. Obtain the neperian logarithm to each of the values ordered in step 1.
7. Order Excel to fit a regression model:
   On the tool ribbon, click:

   Data / Data Analysis / Regression

   Select the input Y-rank = double logarithm column.

   Input X-rank = column of the logarithm of ordered values.
   Choose the output area.

8. Observe in the results the Coefficients Intercept and Variable X1. The coefficient Variable X1 corresponds to the shape parameter of the distribution. The scale parameter of the Weibull is obtained by exponentiating the negative quotient between the coefficients Intercept and Variable X1.

	A	B	C	D	E	F	G	H
	Car (index)	Duration (months) in ascending order	Median Rank MR $=(i-0.3)/(99+0.4)$	$\dfrac{1}{1-MR}$	$LN(LN(\frac{1}{1-MR}))$ Y dependent variable	X Regressor Variable LN(Duration)	=LN(B2)	
1	i							
2	1	2.2	0.007	1.007	-4.952	0.788	Regression Statistics	
3	2	3	0.017	1.017	-4.060	1.099		Coefficients
4	3	3.3	0.027	1.028	-3.592	1.194	Intercept	-5.33
5	4	3.5	0.037	1.039	-3.272	1.253	X Variable 1	1.80
6	5	3.7	0.047	1.050	-3.027	1.308		
7	6	3.7	0.057	1.061	-2.829	1.308	Weibull Parameters	
8	7	3.9	0.067	1.072	-2.662	1.361	Scale =	1.80
9	8	4	0.077	1.084	-2.518	1.386	Shape =	19.17
10	9	5	0.088	1.096	-2.390	1.609		
11	10	5	0.098	1.108	-2.276	1.609		=EXP(-H4/H5)

FIGURE 6.12   Shape and scale parameters for Weibull.

	A	B	C	D	E	F	G	H	I
1	Min	Max	Range	Sturges: Wide Class I. 1+LOG(99;2)	Scale Beta	Shape Alfa			
2	2.2	50.4	48.2	7.629	19.17	1.8			
3	Duration				=WEIBULL.DIST(C6;$F$2;$E$2;1) - WEIBULL.DIST(C5;$F$2;$E$2;1)				
4	2.2			Class (months)	Observed Frequency Fo	Weibull Probability	Expected Frequency Fe	$\dfrac{(Fe-Fo)^2}{Fe}$	=99*E5
5	3		2.20		1	0.020	1.99	0.4926	
6	3.3	=C5+$D$2	9.83		28	0.239	23.70	0.7784	((F6-D6)^2)/F6
7	3.5		17.46		23	0.311	30.78	1.9678	
8	3.7		25.09		26	0.232	22.99	0.3943	
9	3.7		32.72		11	0.124	12.31	0.1389	
10	3.9		40.35		8	0.051	5.05	1.7254	
11	4		47.98		1	0.017	1.64	0.2487	
12	5		55.61		1	0.005	0.54	0.3981	

FIGURE 6.13   Calculations for the Chi-square test statistic.

From the calculations shown in Figure 6.14, it is found that $\chi2$ is 3.992. The $\chi2$ value is determined from Excel for 0.05 and $v = 5 - 1 - 2 = 2$ degrees of freedom: =CHISQ.INV.RT(0.05;2) = 5.99. Since χ^2 = 3.992 is less than theoretical χ^2 = 5.99, the null hypothesis is not rejected; so it can be concluded that a Weibull distribution of parameters 19.17 and 1.80 adequately models the failure times of the cars under consideration.

	B	C	D	E	F	G
14		Class (months)	Observed Frequency Fo	Weibull Probability	Expected Frequency Fe	$\dfrac{(Fe - Fo)^2}{Fe}$
15		9.83	29	0.260	25.70	0.424
16		17.46	23	0.311	30.78	1.968
17		25.09	26	0.232	22.99	0.395
18		32.72	11	0.124	12.31	0.139
19		40.35	10	0.073	7.22	1.066
20			99	1.0	$\chi^2 =$	3.992

FIGURE 6.14   Grouping classes.

6.14 KOLMOGOROV-SMIRNOV FITTING: LOGNORMAL

Example 6.10: Table 6.15 contains the latency times in days from infection to onset of symptoms, of 100 patients infected with salmonella. Use the K-S test to evaluate the possibility of representing the latency values by a Lognormal distribution at the 5% significance level.

TABLE 6.15   Salmonella latency days

70.5	2.8	47.5	8.4	6.1	23.0	27.4	8.5	3.6	16.8
24.7	25.0	75	12.7	12.6	11.7	6.1	5.8	8.3	10.1
5.3	37.7	12.2	6.9	26.5	39.2	1.3	20.3	9.1	60.7
12.4	13.9	74.0	20.2	7.2	1.4	5.6	0.4	118.7	8.7
1.6	3.4	1.5	0.7	15.8	67.5	2.5	1.9	43.9	8.8
16.8	17.5	30.8	0.9	14.7	1.8	73.2	2.8	29.2	24.0
7.3	50.2	2.3	0.1	30.4	5.4	7.0	46.5	37.2	13.8
10.8	5.4	10.6	14.1	16.8	5.4	18.5	1.3	1.8	89.8
4.8	4.4	20.4	11.6	35.4	5.5	46.3	0.8	194.6	5.1
7.4	12.1	23.4	20.4	5.0	49.9	1.5	33.9	6.7	7.4

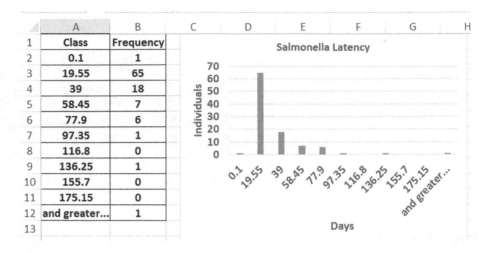

	A	B	C	D	E	F	G	H
1	**Class**	**Frequency**						
2	0.1	1						
3	19.55	65						
4	39	18						
5	58.45	7						
6	77.9	6						
7	97.35	1						
8	116.8	0						
9	136.25	1						
10	155.7	0						
11	175.15	0						
12	and greater...	1						
13								

FIGURE 6.15   Salmonella latency days.

Figure 6.15 suggests the shape of various distributions candidates. Among others probability distributions such as Gamma, Lognormal, Exponential, Weibull. The histogram in Figure 6.15 was obtained through Excel Data Analisys.

The **range** of the data is the difference between the largest value and the smallest: 194.6 − 0.1 = 194.5. Scott's relationship to determine the optimal width **W** of each class interval is:

$$W = 3.49 * s * n^{-1/3},$$

where **s** = 28.236 and **n** = 100.

In the Excel sheet, the calculation of **W** is performed with the following expression:

W = 3.49 * **STDEV.S**(A1:A100)*(100^(−1/3)) = 21.2.

Consequently, the number of classes is obtained by dividing the range by the width: 194.5/21.2. That is, 10 class intervals will be defined. Figure 6.16 presents the values and their logarithms, in addition to the estimation of their parameters.

The critical **D** value obtained from the K-S table is: **D**(0.05;100) = 1.36/10 = 0.136; evidently greater than the maximum difference obtained is 0.0273 (Figure 6.17, cell L12). It is concluded that there is no statistical reason to reject the assumption that the latency times of salmonella conform to a Lognormal distribution of parameters 2.33 and 1.34.

	A	B	C	D	E	F
1	Patient	Latency (Days)	LN (Days)		Lognormal Average:	Lognormal Deviation:
2	1	70.5	4.256		2.329	1.338
3	2	24.7	3.207			
4	3	5.3	1.668		=LN(B2)	
5	4	12.4	2.518			
6	5	1.6	0.470			
7	6	16.8	2.821			
8	7	7.3	1.988			
9	8	10.8	2.380			
10	9	4.8	1.569			
99	98	89.8	4.498			
100	99	5.1	1.629			
101	100	7.4	2.001			

FIGURE 6.16 Obtaining logarithm and parameters of latency data.

	E	F	G	H	I	J	K	L	M	N
1	Lognormal Average:	Lognormal Deviation:		Classes W = 21.2	Observed Frequency	Cumulated Observed Probability	Cumulated Theoretical Probability	Difference		
2	2.329	1.338		21.2	70	0.7	0.706	0.006	=ABS(K2-J2)	
3		=H2+21.2		42.4	15	0.85	0.855	0.005		
4				63.6	7	0.92	0.914	0.006		
5				84.8	5	0.97	0.943	0.027		
6				106	1	0.98	0.959	0.021		
7				127.2	1	0.99	0.970	0.020		
8				148.4	0	0.99	0.977	0.013		
9				169.6	0	0.99	0.982	0.008		
10				190.8	0	0.99	0.986	0.004		
11				212	1	1	0.988	0.012		
12						K-S Maximal Diference D =		0.0273	=MAX(L2:L12)	
13		{=FREQUENCY(B2:B101;H2:H11)}					=LOGNORM.DIST(H3;$E$2;$F$2;1)			
14					=I3/100+J2					

FIGURE 6.17 Calculations of the K-S test.

6.15 GAMMA FITTING USING KOLMOGOROV-SMIRNOV

6.15.1 Thom's Estimator Method

Example 6.12: The additional months of life after a complicated surgical intervention were recorded from a sample of 195 patients. It is required to determine whether the months of survival of a patient can be represented by a Gamma distribution. The survival times are:

52.2	80.9	26.7	77.4	38.2	53.7	49.1	87.4	73.1	77.9	60.3	23.6	8.8
52.9	9.4	76.6	55.6	50.3	28.3	69.3	26.7	29.3	32.2	77.1	40	33.2
32.0	26.8	34.3	49.1	68.0	40.3	36.2	47.4	31.4	53.3	57.9	59.3	91.5
29.0	46.4	76.1	42.4	19.4	126.1	47.7	111.3	32	39.4	38	54.6	15.8
52.0	32.9	67.4	59.6	48.1	28.8	49.2	60.3	9.5	57.2	80.3	66.9	42.6
32.4	58.3	56.7	28.3	52.9	78.9	82.3	75.5	37.3	20	70.7	46.3	36.5
28.6	37.7	57.2	37.7	73.8	79.4	47.7	53.9	74.3	20.3	54.6	46.8	36.4
80.1	77.5	51.1	47.9	45.8	47.4	35.2	98.5	24.1	38.6	42.9	53.8	62.5
70.8	41.9	22.1	97.3	102.5	44.8	127	70.8	12.4	66.4	86.0	25.7	74.6
35.8	43.7	57.2	69.1	68.0	42.5	40.4	72.1	21.1	28.0	47.6	74.7	21.5
13.1	40.3	9.8	30.3	52.0	58.0	37.0	40.0	43.5	50.2	71.8	27.9	27.0
39.6	23.3	76.7	59.5	54.1	44.0	39.6	38.2	52.9	53.7	21.8	43.4	46.3
85.6	60.4	43.9	78.8	85.5	97.7	59.4	33.7	22.3	51.5	42.4	44.3	33.9
68.6	61.1	103.8	40.9	59.7	51.2	34.1	21.2	28.3	43.0	86.2	40.0	83.7
29.4	54.7	79.7	101.1	59.8	41.0	61.0	36.8	44.6	56.5	51.2	45.2	45.9

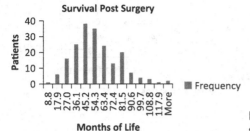

FIGURE 6.18 Histogram of survival after surgery.

Solution

Data values will be placed in Excel range **B2:B196** and its Neperian logarithm in range **C2:C196**. The 195 data vary between a minimum of 8.8 and a maximum of 127. According to Scott's formula: **$3.49 * s * n^{-1/3}$**, for **s** = 13.5 and **n** = 195, the months of life were grouped into eight classes of amplitude 9.

Figure 6.18 shows a histogram containing a shape that may well fit a Gamma distribution.

To estimate its alpha and beta parameters, the Thom's estimator will be used (Shenton and Bowman 1970).

$$\hat{\alpha} = \frac{1}{4 * A}\left(1 + \sqrt{1 + \frac{4 * A}{3}}\right)$$

$$\hat{\beta} = \frac{\bar{x}}{\hat{\alpha}}$$

For this, the value **A** obtained from the neperian logarithm of the average of the **n** = 195 values minus the summation of the logarithms to each life value, divided by the total values is required. That is:

$$A = LN(\bar{x}) - \frac{\sum_1^n LN(x_i)}{n} = 0.106$$

The value **A** = 0.106 (Figure 6.19, cell N5) allows us to estimate the shape parameter **Alpha** = 4.89 through the formula:

$$\hat{\alpha} = \frac{1}{4 * A}\left(1 + \sqrt{1 + \frac{4 * A}{3}}\right) = 4.89$$

And once the shape parameter has been calculated, the scale parameter **Beta** is determined by the formula.

$$\hat{\beta} = \frac{\bar{x}}{\hat{\alpha}} = 10.5$$

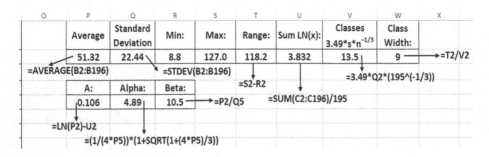

FIGURE 6.19   Statistics values for data fitting.

To evaluate the fit, 13 class intervals will be arranged, with a width of 9 (Figure 6.19).
Then, with this parameter the hypotheses for the goodness-of-fit test are defined as:

H_o: The post-surgery life time of a patient can be represented according to a Gamma
(4.89;10.5) distribution.

H_1: The life span after surgery of a patient cannot be represented as a Gamma
(4.89;10.5) distribution.

Figure 6.20 offers three ways to estimate the parameters of the Gamma distribution,
which show an acceptable similarity of results.

Figure 6.21 shows the calculations of the statistic for the K-S test.

Since in Figure 6.21 there are several classes with less than five individuals, it is
necessary to group some classes, as shown in Figure 6.22.

The test statistic is the maximum value D, therefore $D_{max} = 0.0364$. The value to
compare from the K-S table is:

	A	B	C	D	E	F	G	H
	Pacient	Months of Life					1. Thom Shenton	
1	i	Xi	LN(Xi)	Xi * LN(Xi)				
2	1	8.8	2.17	19.14 →=B2*C2		Average(Xi):	51.32	→ =AVERAGE(B2:B196)
3	2	33.2	3.50	116.28		LN(Average(Xi)):	3.94	→ =LN(G2)
4	3	91.5	4.52	413.25		Sum (LN(Xi))/n:	747.27	→ =SUM(C2:C196)
5	4	15.8	2.76	43.61		A:	0.106	→ =G3-(G4/195)
6	5	42.6	3.75	159.83				
7	6	36.5	3.60	131.30		Shape Alfa=	4.89	→ =(1/(4*G5))*(1+SQRT(1+(4*G5)/3))
8	7	36.4	3.59	130.84		Scale Beta =	10.50	→ =AVERAGE(B2:B196)
9	8	62.5	4.14	258.45				
10	9	74.6	4.31	321.69		2. Ye and Chen		
11	10	21.5	3.07	65.96		Sum (Xi)	10,006.7	→ =SUM(B2:B196)
12	11	27.0	3.30	88.99		Sum LN(Xi)	747.27	→ =SUM(C2:C196)
13	12	46.3	3.84	177.57		Sum (Xi * LN(Xi))	40,356.1	→ =SUM(C2:C196)
14	13	33.9	3.52	119.44				
15	14	83.7	4.43	370.56		Shape Alfa =	4.98	→ =(195*G11)/(195*G13-(G12*G11))
16	15	45.9	3.83	175.63		Scale Beta =	10.30	→ =(1/195^2)*((195*G13)-(G12*G11))
17	16	23.6	3.16	74.61				
18	17	40.0	3.69	147.56		3. Straight from your stats.		
19	18	59.3	4.08	242.10		Shape Alfa	Scale Beta	→ =G2^2/VAR.S(B2:B196)
20	19	54.6	4.00	218.40		5.23	9.81	→ =VAR.S(B2:B196)/G2

FIGURE 6.20   Three methods to estimate the gamma parameters.

	I	J	K	L	M	N	C
1		Months	Observed Frequency	Observed Frequency/195	Gamma Probability: GAMMA.DIST;4.89;10.5;1)	D	
2		8.8	1	0.005128	0.002127	0.0030	
3		17.8	6	0.030769	0.031415	0.0006	
4		26.8	16	0.082051	0.093785	0.0117	
5		35.8	25	0.128205	0.148006	0.0198	
6		44.8	38	0.194872	0.168285	0.0266	
7		53.8	33	0.169231	0.156991	0.0122	
8		62.8	26	0.133333	0.128341	0.0050	
9		71.8	12	0.061538	0.095482	0.0339	
10		80.8	20	0.102564	0.066192	0.0364	
11		89.8	8	0.041026	0.043437	0.0024	
12		98.8	4	0.020513	0.027281	0.0068	
13		107.8	3	0.015385	0.016531	0.0011	
14		116.8	1	0.005128	0.009723	0.0046	
15		125.8	2	0.010256	0.005576	0.0047	
16		{=FREQUENCY(B2:B196;J2:J14)}	=K2/195		=GAMMA.DIST(J2;4.89; 10.5;1)		

=GAMMA.DIST(J3;4.89; 10.5;1)-GAMMA.DIST(J2;4.89; 10.5;1)   =ABS(L2-M2)

FIGURE 6.21   Gamma fit using K-S.

Months	Observed Frequency	Observed Frequency/195	Gamma Probability: GAMMA.DIST;4.89;10.5;1)	D
17.8	7	0.035897	0.033542	0.0036
26.8	16	0.082051	0.093785	0.0117
35.8	25	0.128205	0.148006	0.0198
44.8	38	0.194872	0.168285	0.0266
53.8	33	0.169231	0.156991	0.0122
62.8	26	0.133333	0.128341	0.0050
71.8	12	0.061538	0.095482	0.0339
80.8	20	0.102564	0.066192	0.0364
89.8	8	0.041026	0.043437	0.0024
98.8	10	0.051282	0.059111	0.0172
			Dmax :	0.0364
			D 1.36/SQRT(195) =	0.0974

FIGURE 6.22   Gamma fit after class regrouping.

$$D_{T(0.05;195)} = \frac{1.36}{\sqrt{n}}$$

$D_{T(0.05;195)}$ = 0.0974. Since DT exceeds the D statistic, there is no evidence to reject Ho. In conclusion, the survival times to surgery of the patients can be represented by a Gamma distribution of parameters shape α = 4.89 and scale β =10.5.

The expressions for calculating Beta scale parameter can be simplified by the following relationship between averages of x, LN(x) and x*LN(x):

$$\beta = \frac{N}{N-1} * (\overline{x * LN(x)} - \bar{x} * \overline{LN(x)})$$

For this example, the Excel expression for β would be:

$$= \frac{195}{194} * (\text{AVERAGE}(D2: D196)$$

$$- \text{AVERAGE}(B2: B196) * LN(\text{AVERAGE}(B2: B196)))$$

6.15.2 Ye and Chen Method

Another alternative to estimate the Gamma parameters is offered by Ye and Chen (s.f.). This procedure requires calculating three sums: sum of the x_i values, sum of the natural logarithm of the x_i values, and sum of the product $x_i * LN(x_i)$. The estimation of the parameters $\hat{\alpha}$ and $\hat{\beta}$ is achieved by the following relations:

$$\hat{\alpha} = \frac{1}{n^2}\left(n \sum x_i * LN(x_i)\right) - \sum LN(x_i) * \sum x_i \tag{6.1}$$

$$\hat{\beta} = \frac{n \sum x_i}{n * \sum x_i * Ln(x_i) - \sum Ln(x_i) * \sum x_i} \tag{6.2}$$

Figure 6.23 only shows the first 5 of 195 values. Substituting all 195 survival values into formulas (6.1) and (6.2) the parameters of the Gumbel distribution are obtained: $\hat{\alpha} = 4.98$ and $\hat{\beta} = 10.3$.

In addition to the two methods indicated above, the average and variance formulas can be used to estimate with slightly less accuracy both alpha-form and beta-scale parameters.

$$\text{shape } \alpha = \frac{(\text{mean}(x))^2}{\text{variance}(x)} = 5.23 \text{ and scale } \beta = \frac{\text{mean}(x)}{\text{variance}(x)} = 9.81$$

X_i	Ln(x_i)	X_i * Ln (x_i)
8.8	2.1747	19.1378
33.2	3.5025	116.2846
91.5	4.5163	413.245
15.8	2.76	43.608 1
42.6	3.7518	159.829

FIGURE 6.23   Estimate $\hat{\alpha}$ and $\hat{\beta}$.

6.16 CONCLUSIONS

In this chapter, the classical fitting methods have been presented, recognized to fit a series of values to formal distributions. Usually, more than one distribution can fit the same data set. Therefore, it is advisable to test various distribution models and compare their best suitability. The procedures presented assume the parameters of the distributions as constant rate, invariant over time; for example, a rate of arrival at an ATM service. However, in reality the arrival rate may vary depending on the time of day or the day of the week.

Similarly, procedures Chi-square and K-S fit unimodal input data. Furthermore, the fitted distributions will be used in simulation experiments that will run for long periods of time. These limitations can be managed as required. For example, for distributions with more than one mode, multiple distributions applied to the different sections of the series of observed values can be defined.

In the case of processes where the rate of occurrence changes over time, Leemis (1991) developed a method to define a piecewise constant arrival function $f(t)$ according to distinguishable patterns of arrival. Other modern approaches offer methods to improve the selection of the best probability distribution model to fit, among others, the Akaike Information Criteria (AIC) and Bayesian Information Criteria (BIC). These procedures manage a balance between the goodness of fit and the complexity of the model, in terms of the number of parameters included. In addition, they allow the calculation of indices to select between several candidate models (Burnham et al. 2004).

Statistical Hypothesis Testing

7

7.1 STATISTICAL TESTING IN SIMULATION

The simulation experiment involves numerous assertions about the results arising from the model and the system under study. Sometimes it is necessary (1) to make comparisons between characteristic values of two or more simulation runs; (2) between results of two versions of the model; and (3) between results of the simulation model and values of the real process.

A **hypothesis** is a statement or asseveration. A **statistical hypothesis** is an assumption about the parameters of one or more populations. **Hypothesis testing** is a procedure that is applied to the statement of the hypothesis with the objective of deciding whether or not it is convincing. Testing a hypothesis involves (1) taking a random sample. (2) calculate a test statistic from the sample data. (3) Compare the statistic with the expected tabulated value. (4) making a decision about the hypothesis nule.

Inferential statistics offers methods to estimate parameters and perform hypothesis testing based on the simulation results obtained from the samples observed when running the model.

Grounded on the sample of values extracted from the population, a hypothesis test allows to determine whether a statement about a population parameter is rejected or not.

Some inferential statistical analysis procedures are based on the assumption that the values from the population to be studied can be represented by a normal distribution, characterized by its mean and variance parameters. Generally in the simulation experiment, the shape of the distribution and its parameters are unknown. Moreover, the available data are rarely molded to the idealization of a Gaussian distribution. Consequently, statistical methods based on the assumption of data fitting a normal distribution are inadequate when applied without adherence to the theoretical postulates. This possible duality divides the inferential procedures into parametric and nonparametric.

The so-called nonparametric methods are adaptable to make inference when the data analyzed do not meet the required assumptions of normality shape of the distribution. Since nonparametric methods require fewer assumptions, they should be more widely used and easier to apply. Although the use of nonparametric methods may require larger sample sizes, in practice, expecting data to fit a normal distribution may become a pipe

126

DOI: 10.1201/9781032701554-7

dream; therefore, in the experimental setting of simulation, hypothesis testing prefers nonparametric approaches.

Nonparametric methods define the ordering of observations in order statistics. In a simulation study, it may be necessary to analyze the difference of means between two or more populations or the difference of two or more proportions.

In the case of comparison of means between populations, the parametric test analysis of variance (ANOVA) is a procedure that assumes three or more samples of independent values of normal distribution and equal variance. An alternative nonparametric method is Kruskall Wallis. This procedure only requires samples with independent random values, although they can be ordered according to an ordinal scale.

To compare proportions, the parametric test of comparison of proportions and the nonparametric binomial test are available. Both tests require independent random samples, although the parametric test requires values assimilable to normal distributions.

7.2 PARAMETRIC HYPOTHESIS TESTS

Parametric tests require assumptions about the nature or shape of the populations from which samples are drawn. They are called parametric because their calculation involves the estimation of population parameters from statistical samples. Their application requires a theoretical assumption that the populations follow a normal distribution. In some tests, the variances must be equal and the populations must be independent. Tests for useful verifications applicable in simulation experiments are detailed below.

7.2.1 Hypothesis Test for the Mean

When running a simulation, in each repetition a value or statistic is obtained that reflects the behavior of the variable of interest in the model. The sequence of repetitions produces a random sample, the average of which can be assumed to be a certain constant value. Table 7.1 presents the appropriate procedure for deciding whether the estimated mean represents the population parameter estimated from the results of several repetitions.

TABLE 7.1 Procedure for mean hypothesis testing

Step 1: Define the hypotheses to be tested as:

H_0: $\mu = \mu_0$

H_a: $\mu \neq \mu_0$ **or** $\mu < \mu_0$ **or** $\mu > \mu_0$

Step 2: Define the significance level $\propto$, the sample size **n**, the standard deviation σ and the sample mean $\bar{x}$.

Step 3. Calculate the test statistic:

$$Z = \frac{\bar{x} - \mu_0}{\frac{\sigma}{\sqrt{n}}}$$

(Continued)

TABLE 7.1 (Continued)   Procedure for mean hypothesis testing

Step 4: Establish the rejection criterion for the null hypothesis.

If **Ha**: $\mu_1 \neq \mu_0$ the contrast is bilateral. Calculate $Z_{\frac{\alpha}{2}}$ according to the normal table.

If $Z_{\frac{\alpha}{2}} < Z < Z_{\frac{\alpha}{2}}$ the null hypothesis is not rejected.

If **Ha**: $\mu_1 < \mu_0$ the contrast is one-sided. Calculate Z_{α} according to the normal table. If $Z > -Z_{\alpha}$ the null hypothesis is not rejected.

If **Ha**: $\mu_1 > \mu_0$ the contrast is one-sided. Calculate Z_{α} according to the normal table. If $Z < Z_{\alpha}$ the null hypothesis is not rejected.

Example 7.1: Dispensing Machine

In a simulation study, an engineer obtained data on a 200-liter disinfectant barrel dispensing machine with an average of eight barrels per hour. To validate the average obtained from the simulation, a sample was taken from the existing system for 30 hours and the number of containers dispensed in each hour was recorded. Table 7.2 contains the data. Once its normality condition has been evaluated, it is desired to perform the contrast with a confidence level of 95%, to determine if it is reasonable to assume the average filling rate of eight barrels per hour.

TABLE 7.2   Barrels of disinfectant per hour

7	6	8	5	7	6	3	4	8	7
5	6	8	6	5	4	6	3	6	7
3	4	6	8	9	6	8	7	8	6

Solution

State the hypotheses as follows:

   Ho: $\mu = 8$ barrels/hour

   Ha: $\mu \neq 8$ barrels/hour

Take the data to an Excel sheet. The range **A2:C11** contains the 30 values (Figure 7.1). In cells **E2, E3,** and **E4** calculate the average, standard deviation, and test statistic for the data.

Once the test statistic is obtained determine the critical or rejection region of **Ho**. For this, the significance level α is used. In a bilateral test, the value of $Z_{\frac{\alpha}{2}}$ is sought through the Excel formula:

   =NORM.INV (0.025; 0; 0; 1)

which is arranged in cell **E5,** as shown in Figure 7.1. Then the test statistic should be checked against the region set by: $-Z_{\frac{\alpha}{2}} < Z < Z_{\frac{\alpha}{2}}$. That is, determine whether $Z = -6.38$ is within the interval of values [−1.96 and 1.96].

	A	B	C	D	E
1	**Barrels/Hour**				
2	7	5	3	Average:	6.07
3	6	4	8	Standard Deviation:	1.66
4	6	5	6	Test Statistic Z:	-6.38
5	7	5	9	Z Table Value:	-1.96
6	4	6	3		
7	8	4	3	=(E2-8)/(E3/SQRT(30))	
8	8	6	8		
9	7	6	7	=NORM.INV(0.025; 0; 1)	
10	6	8	8		
11	6	6	7		

FIGURE 7.1 Tabulated data in Excel sheet.

In this case, the test statistic −6.38 is outside the region of non-rejection. Therefore, we conclude the rejection of **Ho**. That is, according to the test, at the 95% confidence level, it cannot be assumed that the average filling rate is eight barrels per hour.

7.2.2 Test for the Difference of Two Averages

We are interested in determining if there is a difference between the averages μ_1 and μ_2 of two populations. To do this, the procedure indicated in Table 7.3 is applied, using both averages obtained in each independent simulation run.

TABLE 7.3 Procedure for testing difference K of two means

Step 1: Define the hypotheses to be contrasted as:

Ho: $\mu_1 - \mu_2 = K$

Ha: $\mu_1 - \mu_2 \neq K$ or $\mu_1 - \mu_2 < K$ or $\mu_1 - \mu_2 > K$

Step 2: Set the significance level α, the sample size **n**, the standard deviations σ_1 y σ_2 and the sample means $\bar{x}_1$ y $\bar{x}_2$

Step 3: Calculate the test statistic with the following formula:

$$Z = \frac{(\bar{x}_1 - \bar{x}_2) - k}{\sqrt{\frac{\sigma_1^2}{n_1} + \frac{\sigma_2^2}{n_2}}}$$

Step 4: Define the rejection region for the null hypothesis.

If **Ha**: $\mu_1 - \mu_2 \neq K$ the contrast is bilateral. Calculate $Z_{\frac{\alpha}{2}}$ according to the normal table. If $Z_{\frac{\alpha}{2}} < Z < Z_{\frac{\alpha}{2}}$ the null hypothesis is not rejected.

If **Ha**: $\mu_1 - \mu_2 < K$ the contrast is one-sided. Calculate Z_α according to the normal table. If $Z > -Z_\alpha$ the null hypothesis is not rejected.

If **Ha**: $\mu_1 - \mu_2 > K$ the contrast is one-sided. Calculate Z_α according to the normal table. If $Z < Z_\alpha$ the null hypothesis is not rejected.

Example 7.2: Difference of Printers

It is desired to know whether a printer of brand A is better than a printer of brand B in terms of ink life in the printing process. To test this hypothesis, the printing process of both machines is simulated to record the daily time to ink exhaustion. For each machine, 20 independent runs were obtained, whose daily duration times are shown in Table 7.4. Perform the comparison of averages with a confidence level of 97%.

TABLE 7.4   Duration of ink

IA	2.3	4.1	4.5	5.4	6	7.3	6	3.4	4.6	5.6
	4.6	5.6	5.7	6	7	6.7	7.8	6.8	7.7	7.8
IB	2.4	5.5	6.7	7	6	5.3	5.5	4.7	4.3	3.4
	3.3	4.4	4.3	5.6	6.7	6.8	4.5	6.7	4	4.3

Solution

Define the hypothesis contrasts as follows:

$$Ho: \mu_A - \mu_B > 0$$

$$Ha: \mu_A - \mu_B \leq 0$$

Arrange the data in an Excel sheet in rows or columns. In this case, the values are placed in rows **1** and **2**. In cells **B6** and **B7**, the population variance of the data is calculated for each printer (Figure 7.2).

From the Excel toolbar run the following sequence:

Data / Data Analysis / z-Test: Two Sample for Means

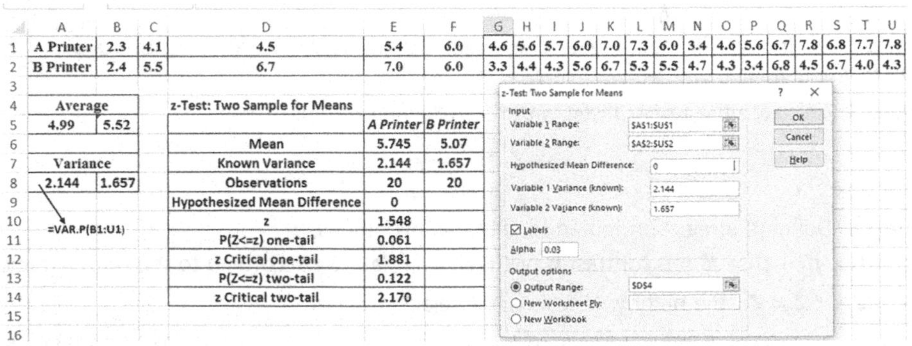

FIGURE 7.2   Contrast for two means.

Figure 7.2 provides the template to be filled in to perform the test and results. The hypothetical difference between the two means is assumed to be 0. Alpha is the significance level for a confidence level of 97%. After clicking OK, the following is obtained.

It shows the test statistic with value of $z = 1.548$, which being a one-sided test, will be compared with the $- Z_\alpha$ from Table 7.5. Therefore, the comparison value is the z for one-tailed: $- Z_\alpha = -1.88$. Since 1.548 is greater than -1.88, there is no evidence to reject the hypothesis that in terms of ink yield printer A is better than printer B.

7.2.3 Test for a Proportion

Table 7.5 presents a useful procedure for testing the value of a hypothesis about the value of a proportion.

TABLE 7.5 Procedure for proportion hypothesis testing

Step 1: Define the hypotheses to be tested as:
 Ho: $p = p_0$
 Ha: $p \neq p_0$ or $p < p_0$ or $p > p_0$

Step 2: Set the significance level α, the sample size **n** and the sample proportions $\hat{p}$ and $\hat{q}$.

Step 3: Calculate the test statistic with the following formula:

$$Z = \frac{\hat{p} - p_0}{\sqrt{\frac{\hat{p}\hat{q}}{n}}}$$

Step 4: Define the rejection region for the null hypothesis.

 If **Ha**: $p \neq p_0$ the contrast is bilateral. Calculate $Z_{\frac{\alpha}{2}}$ according to the normal tab.
 If $-Z_{\frac{\alpha}{2}} < Z < Z_{\frac{\alpha}{2}}$ null hypothesis is not rejected.

 If **Ha**: $p < p_0$ the contrast is one-sided. Calculate Z_α according to the normal table.
 If $Z > -Z_\alpha$ the null hypothesis is not rejected.

 If **Ha**: $p > p_0$ the contrast is one-sided. Calculate Z_α According to the normal table.
 If $Z < Z_{1-\alpha}$ null hypothesis is not rejected.

Example 7.3: Milk Distributor
A milk distributor claims that at least 3 out of 10 milk consumers prefer brand A to brand B. To test this claim, the preference of 30 consumers was simulated and their preference was queried. The data are shown in Table 7.6. Perform the contrast with a confidence level of 95%.

TABLE 7.6 Consumer brand preference

A	B	B	B	A	B	A	B	A	B
A	B	B	B	B	B	A	B	B	B
B	B	B	B	B	B	A	B	B	B

Solution

Define the hypotheses to be tested as:

$Ho: p \geq 0.3$

$Ha: p < 0.3$

Enter the data in the range cells **A2:J4.** Calculate the successes and failures with the function =**COUNTIF**(A2:J4;" "). Calculate the proportions of successes and failures and define the rejection region with the **z** value from the normal table to make the decision to reject or not (Figure 7.3).

	A	B	C	D	E	F	G	H	I	J
1		Brand Preferences								
2	A	B	B	B	A	B	A	B	A	B
3	A	B	B	B	B	B	A	B	B	B
4	B	B	B	B	B	B	A	B	B	B
5										
6	Successes	=COUNTIF(A2:J4;"A")	7							
7	Failures	=COUNTIF(A2:J4;"B")	23							
8	n	30	30							
9	Successes Proportion	=B6/B8	0.2333							
10	Failures Proportion	=B7/B8	0.7666							
11	Hipotetic Value	0.3	0.3							
12	Test Statistic	=(B9-B11)/(SQRT((B9*B10)/B8))	-0.863							
13	Z Table Value	=NORM.S.INV(0.05)	-1.644							

FIGURE 7.3 Results for the contrast.

Figure 7.3 shows the calculations to perform the contrast. The Z-value of the table is calculated with 0.05 because Excel provides the Z-value for the area on the left. Since the test statistic −0.863 in cell C12 is greater than the Z in the table −1.64, there is no evidence to reject the milk distributor's claim. That is, at least 30% of consumers prefer brand A over brand B.

7.2.4 Test for Difference of Proportions

After carrying out the respective repetitions of two independent simulation experiments, the proportion estimates for each population are obtained. Table 7.7 contains a procedure to test the hypothesis about the difference between both proportions from the populations.

TABLE 7.7 Procedure for the difference of proportions

Step 1: Define the hypotheses to be tested as:

Ho: $p_1 - p_2 = K$

Ha: $p_1 - p_2 \neq K$ or $p_1 - p_2 < K$ or $p_1 - p_2 > K$

Step 2: Set the significance level $\propto$, the sample size **n** and the sample proportions and
$\hat{p}_1 \, y \, \hat{p}_2$

Step 3: Calculate the test statistic with the following formula:

$$Z = \frac{(\hat{p}_1 - \hat{p}_2) - k}{\sqrt{\frac{\hat{p}_1 \hat{q}_1}{n_1} + \frac{\hat{p}_2 \hat{q}_2}{n_2}}}$$

Step 4: Define the rejection region for the null hypothesis.

If **Ha**: $p_1 - p_2 \neq K$ the contrast is bilateral. Calculate el $Z_{\frac{\propto}{2}}$ according to the normal tab.

If $-Z_{\frac{\propto}{2}} < Z < Z_{\frac{\propto}{2}}$ null hypothesis is not rejected.

If **Ha**: $p_1 - p_2 < K$ the contrast is one-sided. Calculate $Z_\propto$ according to the normal table.
If $Z > -Z_\propto$ null hypothesis is not rejected.

If **Ha**: $p_1 - p_2 > K$ contrast is one-sided. Calculate el $Z_\propto$ according to the normal table.
If $Z < Z_\propto$ the null hypothesis is not rejected.

Example 7.4: Auto Insurance Claims

An automobile insurance company claims that in a year of coverage men report fewer claims than women. To test this claim, a sample of 15 female and 15 male policyholders was simulated to determine insurance utilization during the previous year. The results are shown in Table 7.8. Perform the contrast at a 97% confidence level.

TABLE 7.8 Use of insurance in any given year

Men	YES	YES	NO	NO	NO	NO	YES	YES
	YES	NO	NO	NO	NO	YES	YES	
Women	NO	NO	YES	YES	YES	YES	NO	
	NO	NO	NO	YES	YES	NO	NO	NO

Solution

Define the hypotheses to be contrasted as:

$$Ho : p_m \leq p_w$$

$$Ha : p_m > p_w$$

Enter the data in columns A and B of the Excel sheet, in cells A2:A16 and B2:B16. Calculate the proportions of successes and failures in each group with the function =COUNTIF(range; criterion). Calculate the test statistic and compare it with the Z value of the normal table, which defines the region of rejection of the null hypothesis (Figure 7.4).

	A	B	C	D	E	F	G
1	Men	Women	Data for Contrast				
2	YES	NO	Men: Claims	7	=COUNTIF(A2:A16;"YES")		
3	YES	NO	Women: Claims	6	=COUNTIF(B2:B16;"YES")		
4	NO	YES	Men: No Claims	8	=COUNTIF(A2:A16;"NO")		
5	NO	YES	Women: No Claims	9			
6	NO	YES	Men Total	15			
7	NO	YES	Women Total	15			
8	YES	NO	Men Claim Proportion	0.467	=D2/D6		
9	YES	NO	Women Claim Proportion	0.400			
10	YES	NO	Men No Claim Proportion	0.533	=D4/D6		
11	NO	NO	Women No Claim Proport.	0.600			
12	NO	YES	Hipotetic Value	0			
13	NO	YES	Test Statistical	0.369			
14	NO	NO	Z Table Value	1.645	=NORM.S.INV(0.9		
15	YES	NO					
16	YES	NO	=(D8-D9)/SQRT(((D8*D10)/D6)+((D9)*D11)/D7)				
17							

FIGURE 7.4 Contrast of proportion of policyholders.

Figure 7.4, cell D13 shows the results of the contrast. Note that for the calculation of the **z** of the table, an area of 0.95 is introduced because it is a one-sided test whose rejection region is on the left. Compare the test statistic of 0.369 with the **z** value of 1.64. Since it is lower, there is no evidence to reject the insurance company's claim. Consequently, it can be asserted that men insured with that company file fewer annual claims than women insured.

7.2.5 Hypothesis Test for Variance

Table 7.9 provides a procedure for hypothesizing the variance of a simulation experiment.

TABLE 7.9   Hypothesis testing for variance

Step 1: Define the hypotheses to be tested as:

Ho: $\sigma^2 = \sigma_0^2$

Ha: $\sigma^2 \neq \sigma_0^2$ or $\sigma^2 < \sigma_0^2$ or $\sigma^2 > \sigma_0^2$

Step 2: Set the significance level $\propto$, sample size **n** and sample deviation $\hat{\sigma}$

Step 3: Calculate the test statistic with the following expression:

$$\chi^2 = \frac{(n-1)\hat{\sigma}^2}{\sigma_0^2}$$

Step 4: Define the rejection region for the null hypothesis. This region is set according to the alternative hypothesis.

TABLE 7.9 (Continued)   Hypothesis testing for variance

If **Ha**: $\sigma^2 \neq \sigma_0^2$ the contrast is bilateral.

Calculate Chi-square value with **n–1** degrees of freedom according to the table. If $\chi^2_{1-\frac{\alpha}{2}} < \chi^2 < \chi^2_{\frac{\alpha}{2}}$ the null hypothesis is not rejected.

If **Ha**: $\sigma^2 < \sigma_0^2$ contrast is one-sided. Calculate $\chi^2_{(1-\alpha)}$ according to the table. If $\chi^2 > \chi^2_{1-\alpha}$ the null hypothesis is not rejected.

If **Ha**: $\sigma^2 > \sigma_0^2$ the contrast is one-sided. Calculate χ^2_α according to the table. If $\chi^2 < \chi^2_\alpha$ the null hypothesis is not rejected.

Example 7.5: Variability in Dispensing Machine

The engineer in Example 7.1 claims that the disinfectant barrel dispensing machine has a variance of less than 2.98. Verify this claim using the data in Table 7.1 for a confidence level of 95%.

Solution

Define the hypotheses to be contrasted as:

Ho: $\sigma^2 < 2.98$ **Ha**: $\sigma^2 \geq 2.98$

Arrange the data in an Excel sheet. Calculate the variance with the formula:

=VAR.P (A2:C11)

Find the Chi-square test statistic in cell **E4**. Obtain the tabulated value corresponding to the one-sided contrast in cell **E5** (Figure 7.5).

Compare the value of the test statistic calculated 25.91 with the Chi-square value in the table: 42.56. Since the value of the test statistic is less than the tabulated Chi, there is no evidence to reject that the amount dispensed per hour has a variance of less than 2.98 barrels.

	A	B	C	D	E
1		**Barrels/Hour**		**Variance:**	2.66
2	7	5	3	**Hypothetical Value:**	2.98
3	6	6	4	**n :**	30
4	8	8	6	**Estimated Chi-Square:**	25.91
5	5	6	8	**Chi-square table:**	42.56
6	7	5	9		
7	6	4	6	=((E3-1)*E1)/E2	
8	3	6	8		
9	4	3	7	=CHISQ.INV.RT(0.05;29)	
10	8	6	8		
11	7	7	6		

FIGURE 7.5 Contrast of variance.

7.2.6 Variance Quotient Hypotheses

Table 7.10 provides a procedure for hypothesizing about the ratio between the variances of two populations in a simulation experiment.

TABLE 7.10 Contrasting hypotheses for variance ratio

Step 1: Define the hypotheses to be contrasted as:

Ho: $\sigma_1^2 = \sigma_2^2$

Ha: $\sigma_1^2 \neq \sigma_2^2$ or $\sigma_1^2 < \sigma_2^2$ or $\sigma_1^2 > \sigma_2^2$

Step 2: Define the significance level α, the size of each sample **n**, and the variance for each population $\hat{\sigma}_1^2$ y $\hat{\sigma}_2^2$

Step 3: Calculate the Fischer test statistic by:

$$\mathcal{F} = \frac{\sigma_1^2}{\sigma_2^2}$$

Step 4: Define the rejection region for the null hypothesis. This region is set according to the alternative hypothesis.

If **Ha**: $\sigma_1^2 \neq \sigma_2^2$ the contrast is bilateral. Calculate the Fischer values according to the table with degrees of freedom for the numerator: $V_1 = n_1 - 1$, and degrees of freedom for the denominator $V_2 = n_2 - 1$.

If $F_{V1,V2,1-\frac{\alpha}{2}} < F < F_{V1,V2,\frac{\alpha}{2}}$ the null hypothesis is not rejected.

If **Ha**: $\sigma_1^2 < \sigma_2^2$ the contrast is one-sided. Calculate the $F_{V1,V2,1-\alpha}$ according to the table.

If $F > F_{V1,V2,1-\alpha}$ the null hypothesis is not rejected.

If **Ha**: $\sigma_1^2 > \sigma_2^2$ the contrast is one-sided. Calculate $F_{V1,V2,\alpha}$ according to the table.

If $F < F_{V1,V2,\alpha}$ the null hypothesis is not rejected.

Example 7.6: Printer Variability
Using the data in Example 7.2, test the hypothesis that the variance of printer A is greater than the variance of printer B. Use a confidence level of 95%.

Solution
Define the hypotheses to be tested as:

Ho: $\sigma_A^2 > \sigma_B^2$ Ha: $\sigma_A^2 \leq \sigma_B^2$

Arrange the data as in Example 7.2 and click on the following sequence:

Data / Data Analysis / F-Test Two-Sample for Variances

Enter the data in the new screen. Note that in the Alpha box 0.95 was enteed because it is a one-sided test to the left. In addition, it is pertinent to clarify that in this test, the Fisher value provided by Excel is that of the area to the left. That is, it is required to calculate $F_{V1,V2,1-\alpha}$ (Figure 7.6).

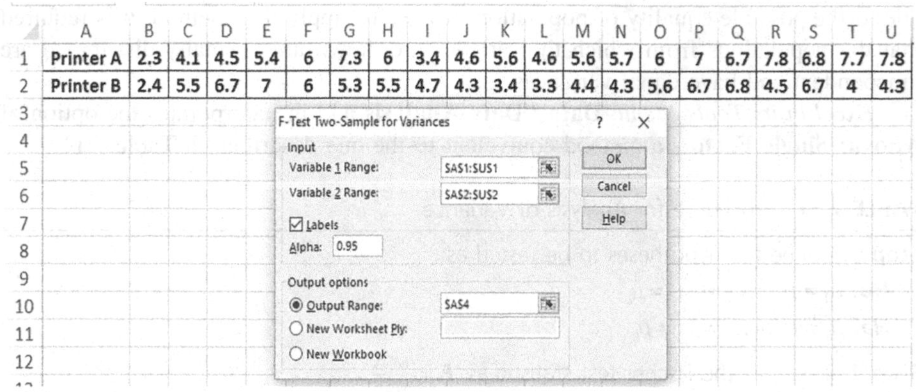

	A	B	C	D	E	F	G	H	I	J	K	L	M	N	O	P	Q	R	S	T	U
1	Printer A	2.3	4.1	4.5	5.4	6	7.3	6	3.4	4.6	5.6	4.6	5.6	5.7	6	7	6.7	7.8	6.8	7.7	7.8
2	Printer B	2.4	5.5	6.7	7	6	5.3	5.5	4.7	4.3	3.4	3.3	4.4	4.3	5.6	6.7	6.8	4.5	6.7	4	4.3

FIGURE 7.6 Contrast for two variances.

Once the data and the Alpha value have been entered, click on OK. The results are shown in Figure 7.7.

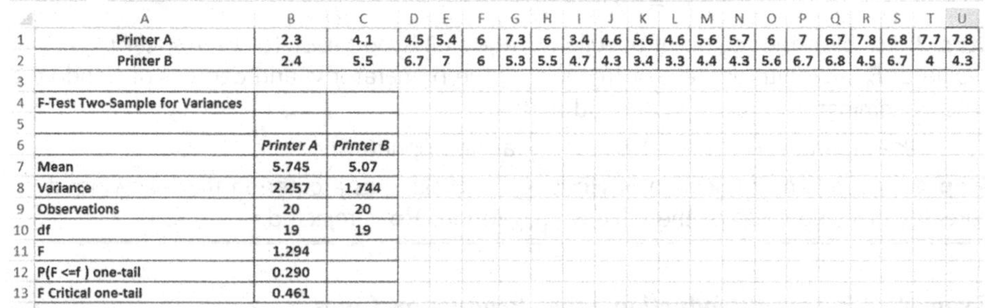

	A	B	C	D	E	F	G	H	I	J	K	L	M	N	O	P	Q	R	S	T	U
1	Printer A	2.3	4.1	4.5	5.4	6	7.3	6	3.4	4.6	5.6	4.6	5.6	5.7	6	7	6.7	7.8	6.8	7.7	7.8
2	Printer B	2.4	5.5	6.7	7	6	5.3	5.5	4.7	4.3	3.4	3.3	4.4	4.3	5.6	6.7	6.8	4.5	6.7	4	4.3
3																					
4	F-Test Two-Sample for Variances																				
5																					
6		*Printer A*	*Printer B*																		
7	Mean	5.745	5.07																		
8	Variance	2.257	1.744																		
9	Observations	20	20																		
10	df	19	19																		
11	F	1.294																			
12	P(F <=f) one-tail	0.290																			
13	F Critical one-tail	0.461																			

FIGURE 7.7 Test results for two variances.

Figure 7.7 shows the calculations for the contrast. The test statistic is **F** = 2.257/ 1.744 = 1.29. For the **Ho** rejection region, the critical value for **F** (one-tailed) is $\mathbf{F}_{0.05;19;19}$ = 0.461. When contrasting the **F** statistic with the critical value, if **F** > $F_{V1,V2,1-\alpha}$ **Ho** is not rejected. Therefore, since 1.29 is greater than 0.461, there is no evidence to reject the hypothesis that the variance of printer A is greater than the variance of printer B.

7.2.7 Hypothesis for Several Means

If it is desired to compare several means, it can be solved using the contrast described in Section 7.2.2, on equality of two means, applied successively to all possible pairs of populations that can be combined with **k** populations. However, this procedure, in addition to being laborious, raises the probability of rejecting the equality of means.

Instead of the above, a method can be applied which under certain conditions allows this type of contrast. The method is known as "Analysis of Variance" or ANOVA. It consists of decomposing the dispersion of the sample data with respect to the mean into two components: one random due to the sampling process and the other due to the possible equality of population means. To apply this method, it is required that the populations from which the samples are drawn are normally distributed are independent and have equal variances.

Excel in its Tools Menu: **Data / Data Analysis** add-in incorporates the option of **Anova: Single Factor**, a method equivalent to the one described in Table 7.11.

TABLE 7.11 Procedure for Analysis of Variance

Step 1: Define the hypotheses to be tested as:

Ho: $\mu_1 = \mu_2 = \mu_3 = \ldots = \mu_k$

Ha: $\mu_1 \neq \mu_2 \neq \mu_3 \neq \ldots \neq \mu_k$

Step 2: Calculate the Fischer test statistic as: $F = \frac{MSA}{MSE}$

Where: $MSA = \frac{SSA}{k-1}$, $MSE = \frac{SSE}{k(n-1)}$ if the sample sizes are equal.

$MSE = \frac{SSE}{N-k}$ with $N = \sum n_i$ If the sample sizes are different.

$$SSA = \sum n_i * (\bar{y}_i - \bar{y})^2 \quad SSE = \sum\sum(\bar{y}_{ij} - \bar{y}_i)$$

$$SST = \sum\sum(y_{ij} - \bar{y})^2$$

Step 3: Establish the region of rejection of the null hypothesis. To do this, calculate the value of $F_{\alpha,v1,v2}$ with degrees of freedom of the numerator v_1 and degrees of freedom of the denominator v_2. Bearing in mind that:

$v_1 = k - 1$ and $v_2 = k(n - 1)$ or $N - k$ as appropriate.

Step 4: The region of rejection of **Ho** is verified taking as a criterion the following: if F is greater than the value of the table $F_{\alpha,v1,v2}$ found, **Ho** is rejected.

Example 7.7 Three Production Lines Production Lines
Table 7.12 shows the results from 10 independent runs of a simulation model for three production lines A, B, and C. The values correspond to the number of failures encountered per line during 1 week. It is desired to know if the average failures of the lines are equal. Use a confidence level of 95%.

TABLE 7.12 Data for production lines A, B, and C

Line A	6	8	10	12	11	13	14	6	8	12
Line B	4	5	8	9	10	11	14	15	12	10
Line C	12	14	6	7	7	8	9	11	13	4

Solution
Enter the data in the Excel sheet. To simplify, we assume normal populations and equal variances.

Define the hypotheses to be tested as:

Ho: $\mu_A = \mu_B = \mu_C$

Ha: $\mu_A \neq \mu_B \neq \mu_C$

	A	B	C	D	E	F	G	H
1		Line		Anova: Single Factor			?	×
2	A	B	C	Input				
3	6	4	12	Input Range:		$AS2:$C$12		OK
4	8	5	14	Grouped By:	⦿ Columns			Cancel
5	10	8	6		○ Rows			Help
6	12	9	7	☑ Labels in first row				
7	11	10	7	Alpha: 0.05				
8	13	11	8	Output options				
9	14	14	9	⦿ Output Range:		E1		
10	6	15	11	○ New Worksheet Ply:				
11	8	12	13	○ New Workbook				
12	12	10	4					

FIGURE 7.8   Data for ANOVA analysis.

From the toolbar follow the following Excel sequence to obtain the ANOVA analysis (Figure 7.8):

Data / Data Analysis / Anova: Single Factor / OK

Once the values have been entered, click OK. Figure 7.9 shows the results of the contrast.

	A	B	C	D	E	F	G	H	I	J	K
1	Line A	Line B	Line C		Anova: Single Factor						
2	6	4	12								
3	8	5	14		SUMMARY						
4	10	8	6		Groups	Count	Sum	Average	Variance		
5	12	9	7		Line A	10	100	10	8.222222		
6	11	10	7		Line B	10	98	9.8	12.4		
7	13	11	8		Line C	10	91	9.1	10.76667	=F.INV.RT(0.05; 2; 27)	
8	14	14	9							=F.INV(0.95; 2; 27)	
9	6	15	11				=SSA		=MSA		
10	8	12	13		ANOVA						
11	12	10	4		Source of Variation	SS	df	MS	F	P-value	F crit
12					Between Groups (treatment)	4.467	2	2.23	0.213	0.81	3.35
13					Within Groups (error)	282.5	27	10.46			
14									=MSA		
15					Total	286.9667	29				
16						=SSE			=Test Statistical		
17											

FIGURE 7.9   Results of the ANOVA application.

To decide whether the hypothesis of equality of means is rejected or not, we take the value of the test statistic, which is 0.213, and compare it with the critical $F_{0.05;2;27;}$ value from the table, whose value is 3.35 (Figure 7.9). If the test statistic is greater than the tabular F value, the null hypothesis is rejected.

Therefore, since 0.213 is less than F critical value 3.35, there is no evidence to reject the null hypothesis, so it can be assumed that the average number of weekly failures occurring in the three production lines are similar.

7.2.8 Test for Several Variances: Bartlett's Test

Bartlett's test is carried out to check that a group of populations have equal variances. In other words, the test checks whether homoscedasticity exists in a group of different populations that follow a normal distribution. The steps to perform the test are shown in Table 7.13.

TABLE 7.13 Contrast of several variances

Step 1: Define the hypotheses to be contrasted as:

Ho: $\sigma_1^2 = \sigma_2^2 = \ldots = \sigma_k^2$

Ha: $\sigma_1^2 \neq \sigma_2^2 \neq \ldots \neq \sigma_k^2$

Step 2: Calculate the Bartlett test statistic as:

$$B = \frac{2.3026 * Q}{h}$$

Where:

$$Q = (N - K) * log_{10} * S_p^2 - \Sigma(n_i - 1) * log_{10} * S_i^2$$

$$S_p^2 = \Sigma_{l=1}^{K} \frac{(n_i - 1)S_i^2}{N - K}$$

$$h = 1 + \left(\frac{1}{3(K-1)} * \left(\Sigma \frac{1}{n_i - 1} - \frac{1}{N-K} \right) \right)$$

Step 3: Establish the region of rejection of the null hypothesis. For this, calculate the value $\chi_{k-1, 1-\alpha}^2$.

Step 4: The region of rejection of **Ho** is verified taking as a criterion the following: if **B** is greater than the table value $\chi_{k-1, 1-\alpha}^2$ **Ho** is rejected.

Example 7.8 Variability in Three Production Lines in 3 Production Lines

Using the data in Example 7.7, verify that the production lines have equal variability with respect to the number of failures encountered in a test week. Use a significance level of 5%.

Solution

Define the hypotheses to be tested as:

$$Ho: \sigma_A^2 = \sigma_B^2 = \sigma_C^2$$

$$Ha: \sigma_A^2 \neq \sigma_B^2 \neq \sigma_C^2$$

Enter the data as in Example 7.7. Although Excel does not provide analyses of various variances, the relationships indicated in the above procedure will be processed. In Figure 7.10, the data and formulas needed to perform the contrast are shown.

	A	B	C	D	E	F
1	Line A	Line B	Line C		S_A^2	=VAR.S(A2:A11)
2	6	4	12		S_B^2	=VAR.S(B2:B11)
3	8	5	14		S_C^2	=VAR.S(C2:C11)
4	10	8	6		n_A	10
5	12	9	7		n_B	10
6	11	10	7		n_C	10
7	13	11	8		S_p^2	=(SUM((F4-1)*F1+(F5-1)*F2+(F6-1)*F3))/F10
8	14	14	9		k	3
9	6	15	11		N	=F4+F5+F6
10	8	12	13		N-k	=F9-F8
11	12	10	4		h	=1+((1/3*(F8-1))*((1/(F4-1))+(1/(F5-1))+(1/(F6-1)+(1/F10))))
12					Q	=F10*LOG10(F7)-(((F4-1)*LOG10(F1))+((F5-1)*LOG10(F2))+((F6-1)*LOG10(F3)))
13					B	=(2.3026*F12)/F11
14					$\chi^2_{1-\alpha;\,k-1}$	=CHISQ.INV(0.95;2)

FIGURE 7.10 Formulas for the three-line test.

Figure 7.11 presents the values of **B** and Chi-square. When comparing both values, since **B** = 0.306 is lower than the Chi-square value = 5.99, the null hypothesis is not rejected. Therefore, it can be concluded with 95% confidence that production lines A, B, and C have equivalent variability.

	A	B	C	D	E	F
1	Line A	Line B	Line C		S_A^2	8.22
2	6	4	12		S_B^2	12.4
3	8	5	14		S_C^2	10.77
4	10	8	6		n_A	10
5	12	9	7		n_B	10
6	11	10	7		n_C	10
7	13	11	8		S_p^2	10.46
8	14	14	9		k	3
9	6	15	11		N	30
10	8	12	13		N-k	27
11	12	10	4		h	1.247
12					Q	0.166
13					B	0.307
14					$\chi^2_{1-\alpha;\,k-1}$	5.991

FIGURE 7.11 Results for three-line contrast A, B, and C.

7.2.9 Proportions Test: Independent Samples

For the application of this test, the populations to be analyzed must be independents. The procedure to be followed is described in Table 7.14.

TABLE 7.14   Contrast for multiple proportions

Step 1: Define the hypotheses to be tested as:

   Ho: $p_1 = p_2 = \ldots = p_k$

   Ha: $p_1 \neq p_2 \neq \ldots \neq p_k$

Step 2: Calculate the Chi-square test statistic as:

$$\chi_P^2 = \frac{1}{\bar{p}\bar{q}} \Sigma\, n_i * (p_i - \bar{p})^2$$

Where:

   $\bar{p}$ = proportion of successes in the total sample

   $\bar{q}$ = proportion of failures in the total sample

   p_i = proportion of successes in each sample

   n_i = size of each sample **i**

Step 3: Establish the region of rejection of the null hypothesis.

   For this, calculate the value: $\chi_{k-1,1-\alpha}^2$.

Step 4: The region of rejection of **Ho** is verified by taking as a criterion the following:

   If χ_P^2 is greater than the table value $\chi_{k-1;\, 1-\alpha}^2$ then **Ho** is rejected.

Example 7.9 Failure of Fiscal Machines

Desired to know whether the failure rate on three fiscal machines is the same. To test this, the three machines are tested for 15 days and **1** is recorded if the machine does not fail during the day and **0** otherwise. The daily operation of the three machines was simulated. The data from the experiment are shown in Table 7.15. To contrast the equivalence of failure proportions use a significance level of 5%.

TABLE 7.15   Daily failure data of the machines

Machine 1	1	1	0	0	1	1	1	0	0	1	1	0	1	0	1
Machine 2	1	0	0	0	1	0	1	0	1	1	1	1	0	0	1
Machine 3	1	1	1	1	0	0	0	1	0	0	1	0	1	0	0

Solution

Define the hypotheses to be tested as:

   Ho: $p_1 = p_2 = p_3$ **Ha**: $p_1 \neq p_2 \neq p_3$

where p_i is the failure rate of machine **i** during the 15 days of testing.

To perform the contrast in Excel, we proceed to enter the data by row or by column. Then count the successes and failures in the total sample, and by machine, in order to calculate the test statistic. The calculations are shown in Figure 7.12.

	A	B	C	D	E	F	G	H	I	J	K	L	M	N	O	P	Q	R	S
1	Machine 1	1	1	0	0	1	1	1	0	0	1	1	0	1	0	1		n	15
2	Machine 2	1	0	0	0	1	0	1	0	1	1	1	1	0	0	1		P_1	=COUNTIF(B1:P1; 0)/\$S\$1
3	Machine 3	1	1	1	1	0	0	0	1	0	0	1	0	1	0	0		P_2	=COUNTIF(B2:P2; 0)/\$S\$1
4																		P_3	=COUNTIF(B3:P3; 0)/\$S\$1
5																		$\bar{P}$	=COUNTIF(B1:P3; 0)/45
6																		$\bar{q}$	=1-S5
7																		χ_p^2	=(1/(S5*S6))*((S1*(S2-S5)^2)+(S1*(S4-S5)^2))
8																		$\chi_{0.95;2}^2$	=CHISQ.INV(0.95; 2)

FIGURE 7.12 Data and formulas for the proportions test.

Once the calculations are obtained, take the value $\chi_p^2 = 0.536$. See Figure 7.13 and compare it with the tabular value $\chi_{0.95;2}^2 = 5.991$. Since 0.536 is less than 5.991, there is no evidence to reject the null hypothesis. Therefore, it cannot be objected that the failure ratio on the three machines is the same.

	A	B	C	D	E	F	G	H	I	J	K	L	M	N	O	P	Q	R	S
1	Machine 1	1	1	0	0	1	1	1	0	0	1	1	0	1	0	1		n	15
2	Machine 2	1	0	0	0	1	0	1	0	1	1	1	1	0	0	1		P_1	0.400
3	Machine 3	1	1	1	1	0	0	0	1	0	0	1	0	1	0	0		P_2	0.467
4																		P_3	0.533
5																		$\bar{P}$	0.467
6																		$\bar{q}$	0.533
7																		χ_p^2	0.536
8																		$\chi_{0.95;2}^2$	5.991

FIGURE 7.13 Results of the proportions contrast.

7.3 NONPARAMETRIC HYPOTHESIS TESTING

Some experiments produce observations that are difficult to quantify; results that while they can be ordered, their location on a measurement scale is arbitrary. In these cases, parametric tests are inadequate. Therefore, **nonparametric methods** are required. Nonparametric tests are also useful for making inferences in situations where there are doubts about the assumptions required in parametric methods, for example, suppositions about normality, homoscedasticity (equality of variances), and independence.

Several investigations have shown that nonparametric tests are almost as capable of detecting differences between populations as parametric tests when the required theoretical assumptions are not met. For this reason, some tests that will be useful for analyzing simulation processes are presented below.

7.3.1 Test of Means: Wilcoxon-Mann-Whitney Test

The Wilcoxon-Mann-Whitney test, also called the Mann-Whitney U test, uses two samples to evaluate whether there is a significant difference between the location parameters of two populations. Specifically, **it compares the averages of two groups** of independent values or in other words, it determines whether or not they come from the same population.

Given two random samples $x_1, x_2, x_3, ..., x_{n1}$ and $y_1, y_2, y_3, ..., y_{n2}$ coming from two populations. It is desired to determine if the values follow the same probability distribution.

To perform this test the procedure described in Table 7.16 is followed.

TABLE 7.16 Wilcoxon-Mann-Whitney Test

Step 1: Define the hypotheses to be tested as:
  Ho: $\mu_1 = \mu_2$
  Ha: $\mu_1 \neq \mu_2$ or $\mu_1 < \mu_2$ or $\mu_1 > \mu_2$

Step 2: Combine the values of the two samples obtained in the simulation, ordering them from lowest to highest. Assign a position or rank: 1, 2, ..., n_1+n_2. If two or more values are equal, each of them should be assigned a position or rank equal to the average of the positions they would occupy.

Step 3: Obtain the sum of the ranks of each sample. These sums are denoted by R_1 and R_2, where n_1 y n_2 are the respective sample sizes. If these are not equal the smaller sample is chosen as n_1.

Step 4: To test the difference between the sums of the ranks, the statistic is used:
  $$U = n_1 n_2 + \frac{n_1(n_1 + 1)}{2} - R_1$$
The sampling distribution **U** is symmetric and has mean and variance given, respectively, by:
  $$\mu_u = \frac{n_1 n_2}{2} \quad \text{and} \quad \sigma_u^2 = \frac{n_1 n_2 (n_1 + n_2 + 1)}{12}$$
If both, n_1 and n_2, are at least equal to 8, then the distribution of **U** is approximately normal, so that:
  $$z = \frac{U - \mu_U}{\sigma_U}$$
has a standard normal distribution. Therefore, the parametric normality test is applied with this statistic, taking into account the alternative hypothesis to define the region of rejection of the null hypothesis.

Example 7.10 Average Stoppages on 12 Machines
Results are available from 10 independent simulation model runs of two production lines A and B. Each production line has six machines. The number of machines that fail weekly on production line A is Poisson averaging 3.6 and on line B also Poisson averaging 4.1. The simulated values correspond to the number of machines stopped weekly. It is desired to know if the average stoppages in both lines are equal. Use a confidence level of 95%. Table 7.17 contains the simulated results.

TABLE 7.17 Machines stopped per week on two production lines

Line A	6	4	5	4	5	2	3	5	3	4
Line B	2	3	1	5	2	4	6	4	3	5

Solution

Define the hypotheses to be tested as:

Ho: $\mu_A = \mu_B$

Ha: $\mu_A \neq \mu_B$

Arrange the data in the Excel sheet by columns A and B. In range A2:B11 place values corresponding to line A; in range A12:A21 place values from line B.

Combine the values of the two samples and order them from smallest to largest. Identify the values by production line.

Copy range A2:B21 to range C2:D21. From tools menu:

SORT / Select range C2:D21 / **Sort** / Smallest to Largest / **OK**

Assign a position or rank, keeping in mind that in repeated values, the rank is the average of the originally assigned positions. To determine ranks, in cell **E2** type:

=RANK.AVG(C2; $C$2:$D$21; 1)

Copy cell E2 until cell E21 (Figure 7.14).

	A	B	C	D	E
1	Machines Broken	Production Line	Sorted Machines Broken	Line	Ranks
2	6	A	1	B	1
3	4	A	2	A	3
4	5	A	2	B	3
5	4	A	2	B	3
6	5	A	3	A	6.5
7	2	A	3	A	6.5
8	3	A	3	B	6.5
9	5	A	3	B	6.5
10	3	A	4	A	11
11	4	A	4	A	11
12	2	B	4	A	11
13	3	B	4	B	11
14	1	B	4	B	11
15	5	B	5	A	16
16	2	B	5	A	16
17	4	B	5	A	16
18	6	B	5	B	16
19	4	B	5	B	16
20	3	B	6	A	19.5
21	5	B	6	B	19.5

FIGURE 7.14 Combining, ordering, and assigning ranges.

Obtain the sum of the ranks for each sample and calculate the test statistic. For the rejection region follow step 4 of the procedure described in Section 7.2.2. In the referenced example, calculate the $Z_{\frac{\alpha}{2}}$ according to the normal table. If $-Z_{\frac{\alpha}{2}} < Z < Z_{\frac{\alpha}{2}}$ the null hypothesis is not rejected (Figures 7.15 and 7.16).

	A	B	C	D	E	F	G	H
1	Machines Broken	Production Line	Machines Broken Ascending Order	Line	Ranks			
2	6	A	1	B	=RANK.AVG(C2;$C$2:$D$21;1)		R1	=SUMIFS(E2:E21;D2:D21;"A")
3	4	A	2	A	=RANK.AVG(C3;$C$2:$D$21;1)		R2	=SUMIFS(E2:E21;D2:D21;"B")
4	5	A	2	B	=RANK.AVG(C4;$C$2:$D$21;1)		n_1	10
5	4	A	2	B	=RANK.AVG(C5;$C$2:$D$21;1)		n_2	10
6	5	A	3	A	=RANK.AVG(C6;$C$2:$D$21;1)		U	=(H4*H5)+(H4*(H4+1)/2)-H2
7	2	A	3	A	=RANK.AVG(C7;$C$2:$D$21;1)		μ_U	=(H4*H5)/2
8	3	A	3	B	=RANK.AVG(C8;$C$2:$D$21;1)		σ_U^2	=((H4*H5)*(H4+H5+1))/12
9	5	A	3	B	=RANK.AVG(C9;$C$2:$D$21;1)		Z	=(H6-H7)/SQRT(H8)
10	3	A	4	A	=RANK.AVG(C10;$C$2:$D$21;1)		$Z_{\alpha/2}$	=NORM.S.INV(0.025)
11	4	A	4	A	=RANK.AVG(C11;$C$2:$D$21;1)			
12	2	B	4	A	=RANK.AVG(C12;$C$2:$D$21;1)			
13	3	B	4	B	=RANK.AVG(C13;$C$2:$D$21;1)			
14	1	B	4	B	=RANK.AVG(C14;$C$2:$D$21;1)			
15	5	B	5	A	=RANK.AVG(C15;$C$2:$D$21;1)			
16	2	B	5	A	=RANK.AVG(C16;$C$2:$D$21;1)			
17	4	B	5	A	=RANK.AVG(C17;$C$2:$D$21;1)			
18	6	B	5	B	=RANK.AVG(C18;$C$2:$D$21;1)			
19	4	B	5	B	=RANK.AVG(C19;$C$2:$D$21;1)			
20	3	B	6	A	=RANK.AVG(C20;$C$2:$D$21;1)			
21	5	B	6	B	=RANK.AVG(C21;$C$2:$D$21;1)			
22								

FIGURE 7.15　Formulas for the contrast.

	A	B	C	D	E	F	G	H
1	Machines Broken	Production Line	Machines Broken Ascending Order	Line	Ranks			
2	6	A	1	B	1		R1	116.5
3	4	A	2	A	3		R2	93.5
4	5	A	2	B	3		n_1	10
5	4	A	2	B	3		n_2	10
6	5	A	3	A	6.5		U	38.5
7	2	A	3	A	6.5		μ_U	50
8	3	A	3	B	6.5		σ_U^2	175
9	5	A	3	B	6.5		Z	-0.869
10	3	A	4	A	11		$Z_{\alpha/2}$	-1.96
11	4	A	4	A	11			
12	2	B	4	A	11			
13	3	B	4	B	11			
14	1	B	4	B	11			
15	5	B	5	A	16			
16	2	B	5	A	16			
17	4	B	5	A	16			
18	6	B	5	B	16			
19	4	B	5	B	16			
20	3	B	6	A	19.5			
21	5	B	6	B	19.5			

FIGURE 7.16　Results for the contrast of means.

Since $Z = -0.869$ falls in the region $(-1.96; 1.96)$ the null hypothesis is not rejected. Therefore, according to Wilcoxon-Mann-Whitney, there is no evidence to challenge the assertion that the averages of machines stopped weekly on both lines are equal.

7.3.2 Kruskall-Wallis Test of Means

The Kruskall-Wallis test is a generalization of the test described above for more than two samples. It is used to test the null hypothesis that **k** independent samples come from identical populations. Introduced in 1952 by W. H. Kruskall and W. A. Wallis, the test is a nonparametric procedure for testing the equality of means in the analysis of variance of a factor, when the assumption that the samples were selected from populations that follow a normal distribution is considered unnecessary. To perform the test the procedure described in Table 7.18 is followed.

TABLE 7.18   Kruskall-Wallis test

Step 1: Define the hypotheses to be tested as:
   Ho: $\mu_1 = \mu_2 = \mu_3 = \ldots \ldots \ldots \ldots = \mu_k$
   Ha: $\mu_1 \neq \mu_2 \neq \mu_3 = \ldots \ldots \ldots \ldots \neq \mu_k$
Step 2: Combine all the values of the **k** samples obtained in the simulation, sorting them from smallest to largest.
Assign a position or rank. If two or more values are equal, each should be assigned a position or rank equal to the average of the positions they would occupy.
Step 3: Obtain the sum of the ranks of each sample. These sums are denoted by
$R_1, R_2, \ldots, R_k$, where $n_1, n_2, \ldots, n_k$ are the respective sample sizes and $n = \sum_{i=1}^{K} n_i$
Step 4: Calculate the test statistic:
$$H = \frac{12}{n*(n+1)} \sum_{i=1}^{k} \frac{R_i^2}{n_i} - 3(n+1)$$
Step 5: The region of rejection of **H₀** is verified by taking the Chi-square distribution as a criterion. The value is looked up in the χ^2 table, for degrees of freedom equal to **k–1** and confidence level **1–α**. If the **H** value is greater than the χ^2 value found then **H₀** is rejected.

Example 7.11: Four Shift Production
In a simulation experiment, 10 results were obtained for each of the four shifts of daily TV set production. It is desired to compare the averages of defective units produced in the independent daily shifts in the factory. The simulation results are shown in Table 7.19. Perform the contrast of means with a confidence level of 97%.

TABLE 7.19 Defective units in the four shifts

Shift 1	3	2	1	4	3	4	3	2	1	1
Shift 2	2	2	1	2	3	3	4	2	3	1
Shift 3	1	3	2	3	2	1	1	2	1	1
Shift 4	2	3	1	3	2	3	1	3	2	4

Solution

Define the hypotheses to be tested as follows:

$$Ho: \mu_1 = \mu_2 = \mu_3 = \mu_4$$

$$Ha: \mu_1 \neq \mu_2 \neq \mu_3 \neq \mu_4$$

Arrange by columns the data in an Excel sheet. Order the values of the entire sample from smallest to largest identifying the values of each sample. Assign ranges following the procedure described in Example 7.10 (Figure 7.17). Initially, all data wrere located at cell range A3:B42.

	A	B	C	D	E	F	G	H	I	J
1	Initial Data			Ordered Data				Ordered Data		
2	Defective Units	Shift		Defective Units	Shift	Ranks		Defective Units	Shift	Ranks
3	3	Shift 1		1	Shift 1	6.5		2	Shift 3	18.5
4	2	Shift 1		1	Shift 1	6.5		2	Shift 4	18.5
5	1	Shift 1		1	Shift 1	6.5		2	Shift 4	18.5
6	4	Shift 1		1	Shift 2	6.5		2	Shift 4	18.5
7	3	Shift 1		1	Shift 2	6.5		3	Shift 1	30.5
8	4	Shift 1		1	Shift 3	6.5		3	Shift 1	30.5
9	3	Shift 1		1	Shift 3	6.5		3	Shift 1	30.5
10	2	Shift 1		1	Shift 3	6.5		3	Shift 2	30.5
11	1	Shift 1		1	Shift 3	6.5		3	Shift 2	30.5
12	1	Shift 1		1	Shift 3	6.5		3	Shift 2	30.5
13	2	Shift 2		1	Shift 4	6.5		3	Shift 3	30.5
14	2	Shift 2		1	Shift 4	6.5		3	Shift 3	30.5
15	1	Shift 2		2	Shift 1	18.5		3	Shift 4	30.5
16	2	Shift 2		2	Shift 1	18.5		3	Shift 4	30.5
17	3	Shift 2		2	Shift 2	18.5		3	Shift 4	30.5
18	3	Shift 2		2	Shift 2	18.5		3	Shift 4	30.5
19	4	Shift 2		2	Shift 2	18.5		4	Shift 1	38.5
20	2	Shift 2		2	Shift 2	18.5		4	Shift 1	38.5
21	3	Shift 2		2	Shift 3	18.5		4	Shift 2	38.5
22	1	Shift 2		2	Shift 3	18.5		4	Shift 4	38.5

FIGURE 7.17 Data and assignment of ranges for Kruskal-Wallis test.

Once the ranges have been assigned calculate the test statistic. To do so, apply the formulas described in Figure 7.18. Column **O** shows the results obtained from the expressions stated in column **N**.

	K	L	M	N
1				
2		R1	=SUMIFS($F$3:$F$42; $E$3:$E$42; "Shift 1")	225
3		R2	=SUMIFS($F$3:$F$42; $E$3:$E$42; "Shift 2")	217
4		R3	=SUMIFS($F$3:$F$42; $E$3:$E$42; "Shift 3")	149
5		R4	=SUMIFS($F$3:$F$42; $E$3:$E$42; "Shift 4")	229
6		n1	10	10
7		n2	10	10
8		n3	10	10
9		n4	10	10
10		n	=SUM(M6:M9)	40
11		H	=((12/(M10*(M10+1)))*(((M2^2)/M6)+((M3^2)/M7)+((M4^2)/M8)+((M5^2)/M9)))-(3*(M10+1))	3.114
12		k	4	4
13		$\chi^2_{1-\alpha;k-1}$	=CHISQ.INV(0.97; 3)	8.947

FIGURE 7.18 Formulas and calculation of the test statistic.

Compare the value of the **H** statistic = 3.114 with the tabulated Chi-square value = 8.947. Since **H** is less than $\chi^2_{0.97;3}$ the null hypothesis is not rejected. Therefore, it can be concluded that the averages of defective units per shift are similar.

7.3.3 Test of Variances: Mood's Test

Random samples drawn from populations with free distributions are available. Let $x_1, x_2, \ldots, x_n$ and $y_1, y_2, \ldots, y_m$ be the chosen samples. It is desired to compare the variances in both samples. For this, the procedure described in Table 7.20 is followed.

TABLE 7.20 Mood test procedure

Step 1: Define the hypotheses to be tested as:

  Ho: $\sigma_1^2 = \sigma_2^2$

  Ha: $\sigma_1^2 \neq \sigma_2^2$ or $\sigma_1^2 < \sigma_2^2$ or $\sigma_1^2 > \sigma_2^2$

Step 2: Order from smallest to largest all the **N** = n + m observations.

  Take **m** the smallest sample. If they have equal size take any **m**.

Step 3: Assign ranks to the **N** observations. Define **R**(y_i) the ranks of the sample **m**.

Step 4: Calculate the test statistic M as follows:

$$M = \sum_{i=1}^{m} \left(R(y_i) - \frac{N+1}{2} \right)^2$$

Step 5: Calculate The values for the mean and variance of M.

$$\mu_M = m\frac{N^2-1}{12} \quad \sigma_M^2 = nm\frac{(N+1)(N^2-4)}{180}$$

For **Ha**: $\sigma_1^2 \neq \sigma_2^2$ reject Ho if: $|M - \mu_M| > Z_{\frac{\alpha}{2}} * \sigma_M$

For **Ha**: $\sigma_1^2 < \sigma_2^2$ reject Ho if: $M > \mu_M + Z_\alpha * \sigma_M$

For **Ha**: $\sigma_1^2 > \sigma_2^2$ reject Ho if: $M < \mu_M - Z_\alpha * \sigma_M$

Example 7.12: Two Machines and One Mechanic

A factory has two machines and one mechanic to service them. The time that machine A lasts in operation before requiring attention from the mechanic is an exponential variable with mean 1 hour, and the time for machine B is an exponential variable with mean 45 minutes. It is believed that the variances of both machines are equal. To test this, 8 days are simulated for each machine. The results, in hours, are shown in Table 7.21. Perform the variance contrast at 95% confidence level.

TABLE 7.21 Hours to require service

Machine A	Machine B
1.23	1.64
0.21	0.28
0.50	0.67
0.19	0.25
0.11	0.15
1.09	1.45
0.33	0.44
0.89	1.19

Solution

Define the hypotheses to be contrasted as:

$$Ho: \sigma_A^2 = \sigma_B^2$$

$$Ha: \sigma_A^2 \neq \sigma_B^2$$

Arrange the data in an Excel sheet by column. Sort the data from lowest to highest, taking care to identify which machine they come from. Assign the ranks to the 16 observations. Once the ranks have been assigned, calculate the **M** statistic mean and variance. Compare the calculated statistic with the rejection region. See the calculation procedure shown in Figures 7.19 and 7.20.

	A	B	C	D	E	F	G	H	I	J
1	Working (hours)	Machine		Machine	Ordered Times	Rank	Machine B $R(y_i)$		Statisticals	Results
2	1.23	A		A	0.11	=RANK.AVG(E2;$E$2:$E$17;1)	=IF(D2="B"; (F2-17)^2; 0)		N	16
3	0.21	A		B	0.15	=RANK.AVG(E3;$E$2:$E$17;1)	=IF(D3="B"; (F3-17)^2; 0)		M	=SUM(G2:G17)
4	0.5	A		A	0.19	=RANK.AVG(E4;$E$2:$E$17;1)	=IF(D4="B"; (F4-17)^2; 0)		μ	=8*((16^2)/12)
5	0.19	A		A	0.21	=RANK.AVG(E5;$E$2:$E$17;1)	=IF(D5="B"; (F5-17)^2; 0)		σ²	=8*8*((17)*((16^2)-4))/180
6	0.11	A		B	0.25	=RANK.AVG(E6;$E$2:$E$17;1)	=IF(D6="B"; (F6-17)^2; 0)		m	8
7	1.09	A		B	0.28	=RANK.AVG(E7;$E$2:$E$17;1)	=IF(D7="B"; (F7-17)^2; 0)		n	8
8	0.33	A		A	0.33	=RANK.AVG(E8;$E$2:$E$17;1)	=IF(D8="B"; (F8-17)^2; 0)		M - μ_M	=J3-J4
9	0.89	A		B	0.44	=RANK.AVG(E9;$E$2:$E$17;1)	=IF(D9="B"; (F9-17)^2; 0)		Z_{α/2}	=NORM.S.INV(0.975)
10	1.64	B		A	0.5	=RANK.AVG(E10;$E$2:$E$17;1)	=IF(D10="B"; (F10-17)^2; 0)		(Z_{α/2})*σ	=J9*J5
11	0.28	B		B	0.67	=RANK.AVG(E11;$E$2:$E$17;1)	=IF(D11="B"; (F11-17)^2; 0)			
12	0.67	B		B	0.89	=RANK.AVG(E12;$E$2:$E$17;1)	=IF(D12="B"; (F12-17)^2; 0)			
13	0.25	B		A	1.09	=RANK.AVG(E13;$E$2:$E$17;1)	=IF(D13="B"; (F13-17)^2; 0)			
14	0.15	B		B	1.19	=RANK.AVG(E14;$E$2:$E$17;1)	=IF(D14="B"; (F14-17)^2; 0)			
15	1.45	B		A	1.23	=RANK.AVG(E15;$E$2:$E$17;1)	=IF(D15="B"; (F15-17)^2; 0)			
16	0.44	B		B	1.45	=RANK.AVG(E16;$E$2:$E$17;1)	=IF(D16="B"; (F16-17)^2; 0)			
17	1.19	B		B	1.64	=RANK.AVG(E17;$E$2:$E$17;1)	=IF(D17="B"; (F17-17)^2; 0)			

FIGURE 7.19 Formulas for the Mood contrast.

	A	B	C	D	E	F	G	H	I	J
1	Working (hours)	Machine		Machine	Ordered Times	Rank	Machine B $R(y_i)$		Statisticals	Results
2	1.23	A		A	0.11	1	0		N	16
3	0.21	A		B	0.15	2	225		M	641
4	0.5	A		A	0.19	3	0		μ	170.7
5	0.19	A		A	0.21	4	0		σ^2	1523.2
6	0.11	A		B	0.25	5	144		m	8
7	1.09	A		B	0.28	6	121		n	8
8	0.33	A		A	0.33	7	0		$M - \mu_M$	470.3
9	0.89	A		B	0.44	8	81		$Z_{\alpha/2}$	1.96
10	1.64	B		A	0.5	9	0		$(Z_{\alpha/2})*\sigma$	2985.4
11	0.28	B		B	0.67	10	49			
12	0.67	B		A	0.89	11	0			
13	0.25	B		A	1.09	12	0			
14	0.15	B		B	1.19	13	16			
15	1.45	B		A	1.23	14	0			
16	0.44	B		B	1.45	15	4			
17	1.19	B		B	1.64	16	1			

FIGURE 7.20   Calculations for the Mood test.

Figure 7.20 shows the results for $M - \mu = 470.3$, and the comparison value: 2,985.4. Since 470.33 is less than 2985.4, the null hypothesis is not rejected. Consequently, there is no evidence to reject the equality of variances of the running times of both machines.

7.3.4 Test of Variances: Conover's Test

When it is desired to compare more than two variances in populations that do not follow the normal distribution, a useful method is the Conover's test. It is used to verify the hypothesis of homogeneity of variances for the variable of interest in various populations. It is a nonparametric, rank-based test that does not require assuming data with a normal distribution. Since the test evaluates the dispersion of the values to their averages, it is applicable to variables defined on an interval scale. That is, the data can be ranked.

We have g groups of data, each group with n_1, n_2, n_3, ..., n_g random data. $X_{i,k}$ corresponds to the ith value of group k and the total of values $N = n_1+n_2+n_3+ ... +n_k$. Table 7.22 details the procedure to be followed to perform Conover's nonparametric test of equality of variances.

TABLE 7.22   Conover test procedure

Step 1: Define the hypotheses to be tested as:

Ho: $\sigma_1 = \sigma_2 = ... = \sigma_g$

The dispersion of the data in the populations is the same. All variances are equal.

Ha: $\sigma_1 \neq \sigma_2 \neq ... \neq \sigma_g$

At least two populations differ in the measure of dispersion of their data.

Where **g** is the number of populations.

(Continued)

TABLE 7.22 (Continued)   Conover test procedure

Step 2: Calculate average $\bar{x}_g$ for each group. Transform one to one value subtracting its group mean under the absolute value formula: $|\bar{x}_{ik} - \bar{x}_k|$ for i = 1 to n_g and k = 1 to g.

Step 3: Sort the transformed data from smallest to largest, taking care to have each value identified by its group. Assign R_{ik} ranks to the total **N** values: R_{ik} = Rank (R_{ik})

Step 4: Obtain the sum of squared ranks for each group. Identify them as $S_1, S_2, ..., S_k$

$$S_k = \sum_{i=1}^{\text{group size}} R_{i,k}^2, \; k = 1, 2, ..., g$$

Step 5: Calculate the test statistic, which has an asymptotic χ^2 distribution with k–1 degrees of freedom.

Calculate:

$$T = \frac{1}{D^2} * \left(\sum_{k=1}^{g} \frac{S_k^2}{n_k} - N\bar{S}^2 \right)$$

Where: $D^2 = \frac{1}{N-1} * \left(\sum_{i=1}^{n_k} \sum_{k=1}^{g} R_{i,k}^4 - N\bar{S}^2 \right)$

R_{ik}: rank assigned to each transformed value.

$$\bar{S} = \frac{1}{N} \sum_{i=1}^{g} S_k$$

Step 6: Calculate the χ^2 value, with k–1 degrees of freedom. If $T > \chi^2$ reject the null hypothesis.

Example 7.13: Variability of Scores

We wish to determine whether there is a difference in the variability of the scores on the first Simulation exam for three different groups. A random sample of 6 students is taken in each class room. The data are shown in Table 7.23.

TABLE 7.23   Scores for the three classes

CLASS 1	CLASS 2	CLASS 3
8	4	2
7	7	4
6	7	5
7	3	9
5	6	6
7	1	3

Solution

Define the hypotheses to be contrasted as:

Ho: $\sigma_1 = \sigma_2 = \sigma_3$

Ha: $\sigma_1 \neq \sigma_2 \neq \sigma_3$

Arrange the data by columns in an Excel sheet (A1:C7) and transform them calculating the average for each group with the Excel function =AVERAGE(); and then for each value data subtract its average and calculate absolute value $|x_i - \bar{x}|$, using the function =**ABS**() (Figure 7.21).

	A	B	C	D	E	F	G	H	I
1		x Class 1	y Class 2	z Class 3	$|x_i - \bar{x}|$	$|y_i - \bar{y}|$	$|z_i - \bar{z}|$		
2		8	4	2	1.333	0.667	2.833	→=ABS(D2-$D$8)	
3		7	7	4	0.333	2.333	0.833		
4		6	7	5	0.667	2.333	0.167		
5		7	3	9	0.333	1.667	4.167		
6		5	6	6	1.667	1.333	1.167		
7		7	1	3	0.333	3.667	1.833		
8	Average:	6.667	4.667	4.833					
9									
10	=AVERAGE(B2:B7)			=ABS(B2-$B$8)	=ABS(C2-$C$8)				

FIGURE 7.21 Data transformation.

Once the data have been transformed, copy ranges B2:B7, C2:C7, and D2:D7 over range J2:J19 and its respective classes in I2:I19.

Sort from smallest to largest using **DATA / Sort** preparing it to assign its respective ranks with =**RANK.AVG**() Excel function (Figure 7.22).

	I	J	K	L	M	N	O	P
1	Class	Order	Rank R_{ik}	$\frac{(Rank)^2}{S_i}$	$(Rank)^4$			
2	Class 3	0.167	1	1	1	→=RANK.AVG(J2; $J$2:$J$19; 1)		
3	Class 1	0.333	3	9	81	→=K3^2		
4	Class 1	0.333	3	9	81	→=K4^4		
5	Class 1	0.333	3	9	81			
6	Class 1	0.667	5.5	30.25	915.06			
7	Class 2	0.667	5.5	30.25	915.06			
8	Class 3	0.833	7	49	2,401			
9	Class 3	1.167	8	64	4,096			
10	Class 1	1.333	9.5	90.25	8,145.06			
11	Class 2	1.333	9.5	90.25	8,145.06			
12	Class 1	1.667	11.5	132.25	17,490.06			
13	Class 2	1.667	11.5	132.25	17,490.06			
14	Class 3	1.833	13	169	28,561			
15	Class 2	2.333	14.5	210.25	44,205.06			
16	Class 2	2.333	14.5	210.25	44,205.06			
17	Class 3	2.833	16	256	65,536			
18	Class 2	3.667	17	289	83,521			
19	Class 3	4.167	18	324	104,976			
20				Sum:	430,845.5			

FIGURE 7.22 Sorting and assigning ranks.

After assigning ranks calculate the test statistic as:

$$T = \frac{1}{D^2} * \left(\sum_{k=1}^{g} \frac{S_k^2}{n_k} - N\bar{S}^2 \right)$$

In Figure 7.23, the calculations and formulas to be used are detailed.

	N	O	P	Q
2		S1	279.75	=SUMIF(I2:I19;"Class 1";L2:L19)
3		S2	962.25	=SUMIF(I2:I19;"Class 2";L2:L19)
4		S3	863	=SUMIF(I2:I19;"Class 3";L2:L19)
5		N	18	
6		$\bar{S}$	116.94	=(P2+P3+P4)/P5
7		$\bar{S}^2$	13,676	=P6^2
8		$\sum R_{ij}^4$	430,845.5	=SUM(M2:M19)
9				
10		D^2	10,863.38	=(P8-(P5*P7))/(P5-1)
11		n1	6	
12		n2	6	
13		n3	6	
14		T	4.17	=(1/P10)*(((P2^2/P11)+(P3^2/P12)+(P4^2/P13))-(P5*P7))

FIGURE 7.23 Results and formulas of Conover's contrast.

The tabulated χ^2 with two degrees of freedom and a 95% confidence level value is 5.99. Our estimated T-statistic is 4.17, as shown in Figure 7.23.

Since **T** $< \chi^2$, according to Conover's test, there is no evidence to reject the claim that there is equality of variances for the grades in the three sections of the simulation course.

Formulation of Basic Simulations

8

8.1 INTRODUCTION

Below is a series of exercises to be solved by simulation using Excel software. The proposed statements, both deterministic and stochastic, are didactic elements, suitable for the flexible spreadsheet environment. They do not correspond to simulations of complex systems, which can be approached by means of specialized simulation languages. The reader is oriented in the methodology of basic models, taking advantage of the facilities offered by a software as widely spread and easy to access as Excel. Since its solution requires vision and individual creativity, each problem poses a solution method among an infinite number of possible options. However, regardless of the logical resolution path selected, your estimates should be acceptably close.

Although for reasons of simplification, in most of the exercises solved here the length of the simulation has not been calculated, nor have confidence intervals been determined for the estimated statistics. The design of the simulation experiment requires that every experimental study includes to determine the size or duration of simulation. Moreover, every estimated statistic must be accompanied by its respective confidence interval. These concepts are covered in Chapter 5, and put into practice in some of the exercises solved.

8.2 DETERMINISTIC SIMULATIONS

Any event whose outcome can be accurately predicted in advance is called deterministic. In deterministic simulation, no input data or variable has random behavior. The deterministic experiment has constant input values and produces constant results. In this approach, the mathematical simulation model expresses the events through completely predictable relationships, and so that, given the same inputs and initial conditions, the model will always yield the same results. By contrast, stochastic simulation experiments contain at least one input value or random variable and each run produces different random results. Since different answers are obtained in each run,

DOI: 10.1201/9781032701554-8

the results of a stochastic model are obtained by averaging the results of several repetitions, whereas the deterministic experiment requires only a single run to achieve accurate answers. In short, constant inputs produce constant outputs and random inputs produce random outputs.

Example 1: Exponential Population Growth

The growth of a population of bacteria can be modeled by an exponential mathematical model. If each bacterium in the population reproduces by binary fission, i.e., it divides into two bacteria in each period, we want to simulate the dynamics of population increase during the next 24 periods. Suppose the initial population is 10 bacteria units. Assume division periods of one hour.

Solution

The model $P_n = P_{n-1}*R$, with $P_0 = 10$, $R = 2$ and $n = 1, 2, ..., 24$, expresses the growth relationship of the bacterial population. The initial population value appears in cell B2 (Figure 8.1). Cell B3 contains the product between the initial population and the growth rate =B2*2. To cover 24 hours of bacterial dynamics, the content of cell B3 is copied into the range B4:B26.

The result of the simulation shows an infection that starts with a population of 10 bacteria. After 24 hours, infection has more than 167 million microbes. The cholera-causing bacteria (Vibrio cholerae) doubles every 17 minutes. In a person an initial population of only 10 bacteria would reach a size of over 2.6 billion in less than 8 hours.

$$Tiempo = \sqrt{2} * Altura/9.81$$

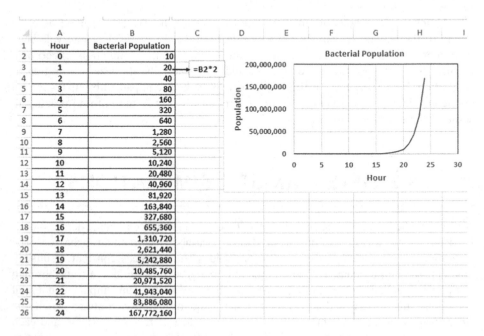

FIGURE 8.1 Evolution of a bacterial population.

	A	B	C	D
	Height	Time		
1	(meters)	(Seconds)		
2	100	4.52	→	=SQRT(2*A2/9.81)
3	200	6.39		
4	300	7.82		
5	400	9.03		
6	500	10.10		
7	600	11.06		
8	700	11.95		
9	800	12.77		
10	900	13.55		
11	1000	14.28		

FIGURE 8.2   Free fall of an object.

Example 2: Free Fall of an Object

When dropping an object from a certain height in free fall, the time in seconds it takes to reach the ground is approximated by the following expression:

$$time = \sqrt{2 * height/9.81}$$

Simulate the fall time for an object dropped from different heights, starting at 100 meters with increments of 100 up to 1,000 meters.

Solution

Figure 8.2 shows the result of the deterministic simulation. The contents of cell B2 are copied to the cell range B3:B11.

Example 3: Monetary Inflation

In a certain country an inflation rate of 5% per year is estimated for the next 10 years. The current rental fee for a house is 100,000 monetary units and is adjusted annually according to the inflation rate. As a deterministic process, (A) Determine the rental value of the house for each of the next 10 years. (B) Simulate the future rental amount for inflation rate variations between 5% and 10% and the time horizon from 1 to 10 years. Use Excel's pivot table facility.

Solution

The Excel finance function:

=**FV**(interest rate; nperiods; payment; [present value]; [typerate])

Obtains the future value of a present value from a constant interest rate, for a given number of periods.

	A	B	C	D
1	Rate:	0.05		
2	Periods:	1		
3	Payment:	100000		
4	Present Value:	105,000.00		
5				
6	Period	Rent		
7	1	105,000.00		=-FV(B1;B2;0;B3)
8	2	110,250.00		
9	3	115,762.50		
10	4	121,550.63		
11	5	127,628.16		
12	6	134,009.56		
13	7	140,710.04		
14	8	147,745.54		
15	9	155,132.82		
16	10	162,889.46		

FIGURE 8.3 Rental value for the next 10 years.

In this case, a pivot table will be created to automate the calculation of the rental value. Initially a single entry table varying the number of payment periods from 1 to 10 years. To declare the variation of a parameter inserted in a table and to place the values per row, the expression is placed in the adjoining cell, just to the left of the first value to be calculated. In case the values are arranged in a column, the formula must be placed in the first cell of the result column (B7). See Figure 8.3.

To obtain the future value of 100,000 varying the number of years from 1 to 10, the function is written in cell B7 as follows:

=–FV(B1; B2; 0; B3)

and apply the procedure described in Table 8.1. The minus sign preceding the name of the function is added to ensure positive values since Excel uses accounting formatting. Thus, the cell will show a positive result. The third argument of the function is left at zero.

TABLE 8.1 Constructing a pivot table from one input in Excel

1. Select the area or range which contains the expression to evaluate and all input values to the table. In this example, select A7:B16

2. In the options bar:
   DATA / What-If Analysis / Data Table ... / Column input cell / **OK**
   Since the number of periods will be varied, indicate the cell reference containing the value to evaluate. In this case, **B2**.

	A	B	C	D	E	F	G	H	I	J	K
1	Rate:	0.05									
2	Periods:	1									
3	Payment:	100000									
4		=-FV(B1;B2;0;B3)									
5											
6	105,000.00	1	2	3	4	5	6	7	8	9	10
7		0.05	105,000.00	110,250.00	115,762.50	121,550.63	127,628.16	134,009.56	140,710.04	147,745.54	155,132.82	162,889.46
8		0.06	106,000.00	112,360.00	119,101.60	126,247.70	133,822.56	141,851.91	150,363.03	159,384.81	168,947.90	179,084.77
9		0.07	107,000.00	114,490.00	122,504.30	131,079.60	140,255.17	150,073.04	160,578.15	171,818.62	183,845.92	196,715.14
10		0.08	108,000.00	116,640.00	125,971.20	136,048.90	146,932.81	158,687.43	171,382.43	185,093.02	199,900.46	215,892.50
11		0.09	109,000.00	118,810.00	129,502.90	141,158.16	153,862.40	167,710.01	182,803.91	199,256.26	217,189.33	236,736.37
12		0.10	110,000.00	121,000.00	133,100.00	146,410.00	161,051.00	177,156.10	194,871.71	214,358.88	235,794.77	259,374.25

FIGURE 8.4 Simultaneous variation of two variables: Interest rate and term.

In Figure 8.3 the range **B7:B16** presents the future value for **n** = 1, 2, … ,10 years of rent.

Figure 8.4 shows the evolution of the rental value for variations of the inflation rate between 5 and 10% (column), and the payment term from 1 to 10 years (row). This output is achieved by defining a two-entry table. A row entry B2, for the values of **n** from 1 to 10 and a column entry B1 for the percentage inflation values.

Cell **A6** contains the function to calculate the future value:

=VF(B1; B2; 0; B3).

To achieve the two-entry table, the procedure described in Table 8.2 is applied.

TABLE 8.2 Creating a two-entry pivot table

1. Select the area or range which contains the expression to be evaluated and all input values to the table. In this example: A6:K12

2. In the options bar:
   DATA / What-If Analysis / Data Table … / Row input cell / Column input cell / **OK**
   Indicate the cell reference containing the row value to be evaluated: **B2**. Then, the cell containing the input column to evaluate: **A2**.

Excel outputs the calculations corresponding to the simultaneous variation of both parameters of the formula and obtains the future value in each of the cases (See Figure 8.4).

Example 4: Drake's Equation: Number of Civilizations
The sun is only one among the 70 sextillion (7×10^{22}) stars in the observable universe. The Milky Way is just one out of 500 billion galaxies in the Cosmos. In such immensity, outside the "ship", a variety of life can exist. The pioneering equation by radio astronomer Frank Drake, 1961, estimates the number of civilizations **C** in the Milky Way galaxy which by means of radio signals might be able to communicate with other living forms:

$$C = R * Fp * Ne * F_L * Fi * Fc * L$$

Where:

R: annual rate of "proper" star formation in the galaxy. (Assumed: 7 stars/year.)

Fp: fraction of stars that have planets in their orbit. (0.5. Half of these stars have planets.)

Ne: number of planets orbiting within the ecosphere of the star; orbits whose distance to the star is neither too close to be too hot nor too far away to be too cold to support life. Assuming each of these stars contains two planets.

F_L: fraction of those planets within the ecosphere on which life has developed. Assume 50% of those planets could develop life.

Fi: fraction of planets on which intelligent life has developed. (0.01. Only 1% would harbor intelligent life.)

Fc: fraction of planets where intelligent beings may have developed technology, and attempt communication. (0.01. Only 1% of such intelligent life can communicate.)

L: time span, measured in years, during which an intelligent, communicating civilization survives until its extinction. (10,000 years. Each civilization would last 10,000 years transmitting signals.)

The British citizen Stephen Hawking, in his book, "The Grand Design", argues that our universe is only one of 10^{50} possible universes. Using the Drake model: (A) For the indicated values simulate and estimate the number of civilizations expected in the Milky Way. (B) Declare a data table to vary in the equation the value of the fractions **Fp** and **FL**, for various values from 0.5 to 1, with increments of 0.1.

Solution
In Figure 8.5, cells B1to B8 contain the values to be included in the Drake equation. The additional term **G** is added by the author. It corresponds to the fraction of sun-like stars; a value of 20% in cell B10 attempts to compensate for the fact that not all stars have sun-like characteristics. The equation is evaluated in cell B9.

	A	B
1	R: annual rate of "proper" star formation	7
2	Fp: fraction of stars that have planets in their orbit	0.5
3	Ne: number of planets orbiting within the ecosphere of the star	2
4	FL: fraction of those planets on which life has developed	0.5
5	Fi: fraction of planets on which intelligent life has developed.	0.01
6	Fc: fraction of planets with intelligent communicators beings	0.01
7	L: time span until its extinction.	15,000
8	G: fraction of type sun stars	0.2
9	Expected amount of civilizations:	1.05
10		
11	=B1*B2*B3*B4*B5*B6*B7*B8	

FIGURE 8.5 Number of expected civilizations in the Milky Way.

To answer question (B), a two-entry table must be formed. For this, arrange the Fp values in cells C9 to H9; and the FL values in cells B10 to B15. To simulate the series of Fp and FL values, declare an Excel table function using the following steps:

1. Select the range B9:H15.
2. In the toolbar select:

Data / What-if Analysis / Data Table ...

The input value to the row is in cell B2 and that of the column in cell B4. Click on **OK**. See Figure 8.6

Figure 8.7 shows the 36 values when evaluating the Drake-I formula.

Example 5: World Population Dynamics
Assume the human life expectancy at 73 years and the annual birth rate at 2.2%. Start the year 2023 with a population of 8 billion people. Simulate the following 100 years of world population to consult: (A) Estimated population for a given year. (B) Year in which the planet will reach a certain population.

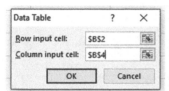

Data Table	?	X
Row input cell:	$B$2	
Column input cell:	$B$4	
OK	Cancel	

FIGURE 8.6   Values for the double-entry table.

	A	B	C	D	E	F	G	H
1	R: annual rate of "proper" star formation	7						
2	Fp: fraction of stars that have planets in their orbit	0.5						
3	Ne: number of planets orbiting within the ecosphere of the st	2						
4	FL: fraction of those planets on which life has developed	0.5						
5	Fi: fraction of planets on which intelligent life has developed	0.01						
6	Fc: fraction of planets with intelligent communicators beings	0.01						
7	L: time span until its extinction.	15,000						
8	G: fraction of type sun stars	0.2	Fp: fraction of stars that have planets					
9	Expected amount of civilizations:	1.05	0.5	0.6	0.7	0.8	0.9	1
10		0.5	1.05	1.26	1.47	1.68	1.89	2.1
11		0.6	1.26	1.51	1.76	2.02	2.27	2.52
12	FL: fraction on which life has developed	0.7	1.47	1.76	2.06	2.35	2.65	2.94
13		0.8	1.68	2.02	2.35	2.69	3.02	3.36
14		0.9	1.89	2.27	2.65	3.02	3.40	3.78
15		1	2.1	2.52	2.94	3.36	3.78	4.2

FIGURE 8.7   Variation of Fp and FL fractions in Drake's Equation.

	A	B	C	D	E	F	G	H	I
1	Initial World Population		Annual Birth Rate	Life Expectancy					
2	8,000,000,000		2.2%	73	A) Year to Search?	2023		B) Size population to search?	Estimated Year:
3					Size Expected Population:	8,066,410,959		10,000,000,000	2049
4									
5		Year	Population at the beginning of the year	Births	Deaths	Population at the end of the year			
6	1	2023	8,000,000,000	176,000,000	109,589,041	8,066,410,959			
7	2	2024	8,066,410,959	177,461,041	110,498,780	8,133,373,220			
8	3	2025	8,133,373,220	178,934,211	111,416,072	8,200,891,359			
9	4	2026	8,200,891,359	180,419,610	112,340,978	8,268,969,991			
10	5	2027	8,268,969,991	181,917,340	113,273,562	8,337,613,770			
11	6	2028	8,337,613,770	183,427,503	114,213,887	8,406,827,385			
12	7	2029	8,406,827,385	184,950,202	115,162,019	8,476,615,569			
13	8	2030	8,476,615,569	186,485,543	116,118,021	8,546,983,090			
14	9	2031	8,546,983,090	188,033,628	117,081,960	8,617,934,758			
15	10	2032	8,617,934,758	189,594,565	118,053,901	8,689,475,422			

FIGURE 8.8   Input and results.

	A	B	C	D	E	F	G	H	I
1	Initial World Population		Annual Birth Rate	Life Expectancy					
2	8000000000		0.022	73	Year to Search?	2023		B) Size	Estimated Year:
3					Size Expected Population:	=LOOKUP(F2;B6:B105;F6:F105)		10000000000	=LOOKUP(H3;C6:C105;B6:B105)
4									
5		Year	Pop at beginning	Births	Deaths	Population at the end of the year			
6	1	2023	=A2	=+C6*$C$2	=+C6/$D$2	=+C6+D6-E6			
7	=A6+1	=B6+1	=+F6	=+C7*$C$2	=+C7/$D$2	=+C7+D7-E7			
8	=A7+1	=B7+1	=+F7	=+C8*$C$2	=+C8/$D$2	=+C8+D8-E8			

FIGURE 8.9   Formulas for world population.

Solution

The input data is entered into the range of cells A2:D2 (Figure 8.8). The format of cell C2 will be defined as percentage and 2.2 will be entered. Once the functions are arranged in row 7, all their content will be copied from row 8 to 105. A partial view of the simulation can be seen in Figure 8.8. It displays the first 10 years of population evolution. The block of cells E2:I3 shows the prediction to a given query.

Figure 8.9 displays the input values and formulas used in this model. The model allows to vary the input data and extend the range of years to be simulated.

Example 6: Linear Regression Forecast

The monthly demand for a product is indicated in the following table

Month	1	2	3	4	5	6	7	8
Units	24,898	47,892	78,321	134,000	158,59	136,700	189,000	235,677

		Monthly Demand										A) Excel Forecast	
	Month X:	1	-2	3	4	5	6	7	8			9	
	Demand Y:	24,898	47,892	78321	134,000	158,590	136,700	189,000	235,677			253,173	
Average X:	4.5			$b = \frac{Sum\ N}{Sum\ D} = \frac{\Sigma(X_i - \bar{X})(Y_i - \bar{Y})}{\Sigma(X_i - \bar{X})^2}$ and $a = \bar{Y} - b\bar{X}$								B) Calculated Forecast	
Average Y:	125,634.8											253,173.32	
X - Mean(X):	-3.5	-2.5	-1.5	-0.5	0.5	1.5	2.5	3.5					
Y - Mean(Y):	-100,737	-77,743	-47,314	8,365	32,955	11,065	63,365	110,042					
[X-Mean(X)]*[Y-Mean(Y)]:	352,579	194,357	70,971	-4,183	16,478	16,598	158,413	385,148					
[X - Mean(X)]^2:	12.3	6.3	2.3	0.3	0.3	2.3	6.3	12.3					
Sum N:	1,190,360												
Sum D:	42												
b =	28,341.9												
a=	-1,903.8												

FIGURE 8.10   Forecast input and output.

Excel includes the FORECAST() function useful for predicting a future value Yn + 1 from two series of available values X1, X2, X3, …, Xn and Y1, Y2, Y3, …, Yn using a linear regression model Y = a + bX. For the input data presented in the table above, estimate the demand for the ninth month using: (A) The Excel FORECAST() function. (B) Applying the regression model through the following expressions:

$$b = \frac{\Sigma(X_i - \bar{X})(Y_i - \bar{Y})}{\Sigma(X_i - \bar{X})^2} \quad \text{and} \quad a = \bar{Y} - b\bar{X}$$

$\bar{X}$ and $\bar{Y}$ correspond to the averages of the series X_i and Y_i, i = 1:n

Solution
The Excel function =FORECAST(9;C3:J3;C2:J2) placed in cell L3 obtains the demand forecast for the ninth month. It only requires as arguments both series of values Yi and Xi, in addition to the value to be predicted Xi + 1.
The procedure to obtain the constants **a** and **b** that fit the linear model begins by calculating the average of each of the series, cells B5 and B6 (Figure 8.10).
The regression equation to approximate the monthly demand is defined according to:

Y = −1,903.8 + 28,341.9 ∗ X

Figure 8.11 presents the formulas used.

$$n = \frac{N1 * N2}{R}$$

Example 7: Capture – Recapture Sampling: Fish Population
The capture-recapture sampling technique is used to estimate the size of populations that are impossible to sample exhaustively. It consists of selecting a group of individuals, marking them and returning them to the population. After a certain period of time, a group of specimens is randomly selected again to determine the fraction of

	A	B	C	D	E	F	G	H	I	J	K	L
1			Monthly Demand									A) Excel Forecast
2		Month X:	1	2	3	4	5	6	7	8		9
3		Demand Y:	24898	47892	78321	134000	158590	136700	189000	235677		=FORECAST(9;C3:J3;C2:J2)
4												
5	Average X:	=AVERAGE(C2:J2)										B) Calculated Forecast
6	Average Y:	=AVERAGE(C3:J3)										=B16 + B15 * L2
7												
8	X - Mean(X):	=C2-$B$5	=D2-$B$5	=E2-$B$5	=F2-$B$5	=G2-$B$5	=H2-$B$5	=I2-$B$5	=J2-$B$5		
9	Y - Mean(Y):	=C3-$B$6	=D3-$B$6	=E3-$B$6	=F3-$B$6	=G3-$B$6	=H3-$B$6	=I3-$B$6	=J3-$B$6		
10	[X-Mean(X)]*[Y-Mean(Y)]:	=B9*B8	=C9*C8	=D9*D8	=E9*E8	=F9*F8	=G9*G8	=H9*H8	=I9*I8		
11	[X - Mean(X)]^2:	=B8*B8	=C8*C8	=D8*D8	=E8*E8	=F8*F8	=G8*G8	=H8*H8	=I8*I8		
12												
13	Sum N:	=SUM(B10:I10)										
14	Sum D:	=SUM(B11:I11)										
15	b =	=B13/B14										
16	a=	=B6-B15*B5										

FIGURE 8.11 Excel forecast formulas.

marked individuals present in the second sample. A basic method is the **Peterson-Lincoln method**. Let N_1: the number of individuals selected and marked the first time; N_2: the number of individuals extracted in the second sample; R: the number of marked individuals obtained in the second selection. The expression:

$$n = \frac{N_1 * N_2}{R}$$

Obtains the estimate of the number of individuals present in the population.

Assume a population of fish in a pond. To estimate its size, 200 fish were initially sampled and tagged. The second sampling yielded 400 fish among which 39 previously tagged fish were recaptured. Simulate to determine the population size assuming recapture values between 35 and 45 individuals.

Solution

Cells B1, B2, and B3 in Figure 8.12 contain the input data 200, 400, and 39, respectively. Cell B4 estimates the population size of individuals according to the Peterson-Lincoln relationship. A single-entry table is constructed to vary the number of recaptured individuals. The range of cells B6:L6 shows the estimated fish population size for each recapture value between 35 and 45 fish. Table 8.3 contains the procedure for developing a single entry table.

	A	B	C	D	E	F	G	H	I	J	K	L
1	Capture 1 and tagged:	200										
2	Capture 2:	400										
3	Recaptured:	39										
4	Estimated population n =	2051	→ =(B1*B2)/B3									
5			35	36	37	38	39	40	41	42	43	44	45
6	=B4 ← 2051	2286	2222	2162	2105	2051	2000	1951	1905	1860	1818	1778

FIGURE 8.12 Fish population for different recapture values.

TABLE 8.3   Steps for constructing a single-entry table

1. Arrange the input data to the table, values 35 to 45 in cells B5 to L5.
2. In cell A6 there must be the reference to cell B4 according to the expression =B4
3. Select the range A5:L6
4. In the toolbar select:

Data / What-if Analysis / Data Table …

Since all values to be varied appear in one row, then Input cell (row) refers to cell B4
Click on **OK**

8.3 STOCHASTIC SIMULATIONS

If the input data or any event of the simulation model contains random components, the simulation is stochastic and its results are also stochastic. This means that in each run of the simulation, the output will present a different result. In general, if the model realization provides dissimilar point estimates, it is advisable to accompany the point estimate with a confidence interval. This requires running the simulation several times. Each run should be independent of the others. It is recommended to perform at least 20 repetitions of the simulation. The average of individual results is an adequate estimator of the parameter, which should be accompanied with its respective confidence interval. Always do it: obtain any estimated value plus its confidence interval.

Example 8: Estimation of π
A method to estimate the value of the constant Pi consists of arranging a circle of radius 1, with center at the origin and in the first quadrant inscribing a square of side 1. Then generate pairs of uniform random values between 0 and 1, corresponding to the coordinates of random points (x, y) on the first quadrant. If the distance from the origin to the point (x, y) is less than 1, it is assumed that the point (x, y) falls inside the circle. The proportion of points inside the circle is an estimate of the fourth part of the value of Pi. Simulate the generation of 50,000 (x, y) points to estimate π.

Solution
After simulating 50,000 points, 39,291 points (x, y) fell inside the circle, then the estimated value of π = 3.14328 appears in cell F8. See Figure 8.13. This result is only a random value. For better accuracy, the simulation can be repeated about 19 times more, and finally, the 20 independent results can be averaged. Always, to obtain a better estimation it is advisable to previously determine the length of the simulation according to the level of confidence and accuracy desired.

Example 9: Rock Paper Scissors Hands-On Game
A game called Rock, Paper, Scissors can be simulated to estimate each bettor's chance of winning. On each bet, on the verbal count: one, two, three, both players at the same

	A	B	C	D	E	F	G
1	Point	coordinate X	coordinate Y	Distance to origin	¿Distance < 1? Yes = 1 Not = 0	Amount of points < 1:	
2	1	0.3752856	0.474675	0.60511	1	39291 —>	=COUNTIF(E2:E50001; 1)
3	2	0.1513369	0.2805261	0.31874	1	=SQRT(B2*B2 + C2*C2)	
4	3	0.0139634	0.5636397	0.56381	1	Value Pi/4:	
5	4	0.9938205	0.6323571	1.17795	0	0.78582 —>	F2/50000
6	5	0.3341748	0.5817872	0.67093	1		
7	6	0.0110623	0.2347805	0.23504	1	Estimated for Pi:	
8	7	0.6079784	0.9565265	1.13339	0	3.14328 —>	=4*F5
9	8	0.7142276	0.2800652	0.76718	1	=RND()	
10	...	0.0706863	0.2510587	0.26082	1		
11	49998	0.7182378	0.4797446	0.86372	1		
12	49999	0.9373942	0.7177081	1.1806	0		
13	50000	0.7961347	0.4246349	0.9023	1 —>	=IF(D50000<1; 1; 0)	

FIGURE 8.13   Estimation of π value.

time show one of three options in their hand: closed hand (rock), open hand (paper), or V-fingers (scissors). If both show the same, there is a tie. If they differ, Rock destroys Scissors, Scissors cuts Paper, or Paper wraps Rock. Simulate the game between two opponents and estimate: (A) Number of bets to simulate to estimate the probability of winning with 1% error and 95% confidence level. (B) Probability of winning for each player.

Solution
In each bet, one of nine possible outcomes occurs. Three of the outcomes favor one player, three favor the opponent, and in three no one wins. Consequently, the theoretical probability of winning a bet is 3/9 = 0.333. The possible outcomes of the game: 1 if bettor 1 wins; 2 if bettor 2 wins; T if both show the same.

1 \ 2	Rock	Paper	Scissor
Rock	T	2	1
Paper	1	T	2
Scissor	2	1	T

In Figure 8.14, cells B5:B14 contain the searching function:

=LOOKUP(lookup value; lookup vector; results vector)

Arranged to mimic the random selection made by player 1. The same function appears in the cells of column C5:C14 to mimic each bet of player 2.

In each bet to determine the winner, each cell in the range D5:D100004 contains the following long nested expression:

	A	B	C	D	E	F	G	H	I	J	K	L
1		Rock	Paper	Scissors		=IF(AND(B5="Rock";C5="Paper");2; IF(AND(B5="Rock";C5="Scissors");1; IF(AND(B5="Paper";C5="Rock");1;					
2	0	0.3333333	0.6666667	1		IF(AND(B5="Scissors";C5="Rock");2; IF(AND(B5="Scissors";C5="Paper");1;"Tie"))))))						
3							Sample size n:					
4	Bet	Player 1	Player 2	Winner		Z(0,025)=	1.96					
5	1	Paper	Paper	Tie		Error =	0.01					
6	2	Paper	Scissors	2		n >=	9604	=(G4*G4)/(4*G5*G5)				
7	3	Scissors	Paper	1								
8	4	Scissors	Rock	2		Times Player 1 Wins:						
9	5	Paper	Paper	Tie		33161	=COUNTIF(D5:D100004;1)					
10	6	Rock	Scissors	1		Times Player 2 Wins:						
11	7	Scissors	Paper	1		33336	=COUNTIF(D5:D100004; 2)					
12	8	Scissors	Paper	1		Probability(Player 1 Wins) =	0.33161	=E9/100000				
13	9	Rock	Scissors	1		Probability(Player 2 Wins) =	0.33336	=E11/100000				
14	10	Scissors	Scissors	Tie								
15	...					=LOOKUP(RAND();$A$2:$D$2;$B$1:$D$1)						

FIGURE 8.14 One hundred thousand bets of rock-paper-scissors.

=IF(AND(B5="Rock"; C5="Paper"); 2; IF(AND(B5="Rock"; C5="Scissors"); 1; IF(AND(B5= "Paper"; C5="Rock"); 1; IF(AND(AND(B5="Paper"; C5="Scissors"); 2; IF(AND(AND(B5= "Scissors"; C5="Stone"); 2; IF(AND(B5="Scissors"; C5="Paper"); 1; "Tie")))))

The probability of success for player 1, with 95% confidence and accuracy of 1%, is obtained by running a simulation length of at least 9,604 bets. See Figure 8.14. However, in this case, 100,000 bets were simulated. The estimated winning probability of player 1 gave 0.33161.

Example 10: Betting on the Roll of a Die and a Coin
In a betting game, the player bets $50 by simultaneously rolling a die and a coin. If he gets six on the die and head on the coin, he wins $500. (A) Determine the number of throws to estimate the expected utility of the gambler with 95% confidence and error of $1. (B) Simulate and estimate the expected utility of a gambler. (C) Establish the value of the prize so that the game is fair.

Solution
A. To define the length of the simulation, the characteristic of the parameter to be estimated is identified in terms of whether it is an average or a proportion. In this case, the parameter to be estimated is the expected utility of the bettor, an average that will be obtained by placing thousands of bets and in the end, averaging the utility. In Chapter 4, the procedure to be followed is described.

Initially it is necessary to observe whether the utility values can be assumed to come from a normal distribution or not. For this purpose, a pilot test of 1,000 bets will be simulated and 1,000 values will be obtained, each corresponding to the utility in each bet.

To assess the normality of the utility the 1,000 values can be taken and processed in any package that performs data fitting to normal probability distributions. In addition, the Excel Data Analysis module can be used to construct a histogram and calculate descriptive statistics. However, it can be observed that the utility variable, for any bettor, only can take two values: minus 50 when he loses and 450 when he wins. Consequently, to calculate the sample size, the central limit theorem does not apply and Chebyshev's formula can be directly used:

$$n \geq \frac{S^2}{\alpha * \in^2}$$

The confidence level is $1 - \alpha = 0.95$, from which $\alpha = 0.05$ is obtained. The error $\in = 1$. The variance S^2 is calculated from a pilot sample of 1,000 bets. See Figure 8.15. Therefore, from the Chebyshev equation, we obtain:

$$n \geq \frac{20,700.5}{0.05 * 1 * 1} = 414,010$$

The sample size is over 414,000 values; a considerable length. It is the price charged by the non-normality of the utility variable.

In Figure 8.15 is a partial view of the simulation in Excel with results for 414,010 bets.

B. The expected simulated utility of the game is $\$-8.35$. It is an unfair game for the gambler.

The theoretical expected value for the utility of the game is calculated as follows:

$$E(\text{Utility per bet}) = (0.5 * 1/6) * 450 - (0.5 * 11/6) * 50 = \$ -8.33.$$

This value -8.33, Figure 8.16, is exact. Compared to the simulated result, it presents a negligible difference. It is the expected loss of a bettor for each bet in that game.

C. To determine the value of the prize so that it is a fair game, it must be simulated trying values greater than the current value of $500 until the average or expected utility $E(U)$ is close to or equal to zero.

	A	B	C	D	E	F	G	H
1	Bet $:	50	Average Utility $ =	-8.34762	→=AVERAGE(D5:D414018)			
2	Prize $:	500	Variance =	20,700.5	→=VAR.S(D5:D1004)			
3								
4	Bet	Coin Side:	Die Side:	Gambler Utility				
5	1	Seal	6	-50	→=IF(AND(B5="Head"; C5=6); $B$2-$B$1; -$B$1)			
6	2	Seal	4	-50				
7	3	Seal	4	-50	Alfa:	0.05		
8	4	Head	6	450	Error:	1	→=D2/(F7*F8*F8)	
9	5	Head	4	-50	Sample size n >=	414,010		
10	6	Head	3	-50				
11	7	Seal	4	-50	→=IF(RAND() < 0,5; "Head"; "Seal")			
12	8	Head	1	-50				
13	9	Seal	1	-50	→=RANDBETWEEN(1 ; 6)			
14	10	Seal	1	-50				
15	11	Seal	2	-50				

FIGURE 8.15 Simulation of the toss of a die and a coin.

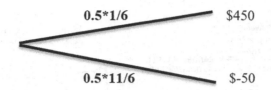

FIGURE 8.16 Possibly outcome in each bet.

	A	B	C	D
1	Bet $:	50	Average Utility $ =	-0.07012
2	Prize $:	601	Variance =	25,390.4
3				
4	Bet	Coin Side:	Die Side:	Gambler Utility
5	1	Head	4	-50
6	2	Seal	1	-50
7	3	Head	1	-50
8	4	Seal	4	-50
9	5	Seal	6	-50
10	6	Head	5	-50
11	7	Head	6	551
12	8	Seal	2	-50
13	9	Seal	5	-50
14	10	Head	6	551
15	11	Seal	1	-50
16	12	Seal	6	-50

FIGURE 8.17 Fair game: Zero expected utility.

This is achieved for a prize, in cell B2, of approximately $601. The result in cell D1 corresponded to a simulated expected utility of $ −0.07012, as shown in Figure 8.17.

Example 11: Law of Large Numbers: Rolling a Die
The roll of a die yields one of six possible outcomes. Theoretically, averaging the sum of the results obtained by rolling a die **N** times is 3.5.

In Figure 8.18, the simulation shows the experimental fulfillment of the law of large numbers. This law states that as the number of values that are averaged increases, the sample average $\bar{X}$ calculated by simulation tends to its expected value μ. If x_1, x_2, x_3, ..., x_n are **n** independent random variables taken from a population, then the sample mean converges to the expected population mean. It occurs as the size **n** increases. That is, the sample mean statistic tends to the expected average parameter. This theorem underlies the estimation procedure applied by the simulation experiment.

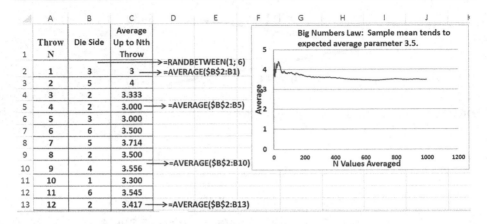

FIGURE 8.18   Throwing a dice: Law of large numbers.

The simulation was performed for 1,000 throws. Column C, from cell C2 to cell C1001, shows the average up to the nth throw. For example, the average in C2 takes only the first throw: =AVERAGE($B$2; B2). However, the average in cell C1001 is calculated as: =AVERAGE($B$2:B1001) averaging all 1,000 values.

Example 12: Law of Large Numbers: Coin Toss
A coin toss can also be used to show the fulfillment of the law of large numbers. In Figure 8.19, the simulation shows the trend of the experiment toward the probability of getting heads.

Example 13: Frequency of Faces in a Die Roll
When a die is tossed, the probability that one of its six sides will appear is equal for each side: 1/6. The following simulates the throwing of a die 100, 1,000, 10,000, and 100,000 times (see Figure 8.20).

The objective is to observe the frequency of occurrence of each possible value and its tendency to be the same, for each of the six sides as the sample size increases.

Once the six possible values have been entered in cells D2 to D7, Figure 8.20, select the range of cells in the column to the right of the values, from E2 to E7, and position the cursor in E2, the first cell selected. The function is written, indicating the data or values from cell B2 to B101; in addition to the grouping values (groups), which are in cells D2 to D7. The following function performs the output grouping of the 6 values.

=FREQUENCY(B2:B101; D2:D7)

It is essential to place the closing parenthesis as follows: press simultaneously the **Ctrl Shift** and **Enter** keys. This is repeated for the values in columns F, G, and H for its corresponding ranges. The Excel frequency function is a matrix function. It therefore requires closing parentheses by pressing the indicated keys at the same time. The results for various values of **N** when a balanced die is rolled, show the increase in

	A	B	C	D	E	F	G	H
1	Throw	Seal (0) Head (1)		N Throws	Heads	Probability of Head		
2	1	0		10	2	0.200000	=E2/10	
3	2	0		50	24	0.480000		
4	3	1		100	49	0.490000		
5	4	0		500	246	0.492000		
6	5	1		1,000	489	0.489000		
7	6	0		5,000	2,458	0.491600		
8	7	0		10,000	4,954	0.495400		
9	8	0		50,000	25,012	0.500240		
10	9	0		100,000	50,081	0.500810	=D10/100000	
11	10	0						
12	11	0		=COUNTIF($B$2:B100001;1)				
13	12	1						
14	13	1		=RANDBETWEEN(0;1)				

FIGURE 8.19 Law of large numbers when tossing a coin.

	A	B	C	D	E	F	G	H
					100 times.	1,000 times.	10,00 times.	100,000 times.
1	Roll	Side		Side	(Expected: 16.7)	(Expected: 166.7)	(Expected: 1,666.7)	(Expected: 16,666.7)
2	1	1		1	20	184	1,728	16,795
3	2	2		2	15	168	1,655	16,566
4	3	4		3	15	150	1,635	16,786
5	4	1		4	16	154	1,676	16,659
6	5	5		5	13	162	1,640	16,693
7	6	6		6	21	182	1,666	16,501
8	7	3						
9	8	6		=FREQUENCY(B2:B101; D2:D7)			=FREQUENCY(B2:B100001; D2:D7)	
10	9	4						
11	10	5		=RANDBETWEEN(1:6)				
12	11	5						

FIGURE 8.20 Frequency on a die toss.

accuracy, matching the number of times each of the possible values is expected to appear.

Example 14: Drawing Two Dice
Simulate the simultaneous throwing of two balanced dice to show graphically the points achieved on each throw.

Solution
Figure 8.21 presents the images resulting from simulating the throwing of both dice. The cells A2 and B2 generate the randomness of each die, by means of the function =RANDBETWEEN(1;6). The range of cells D2:F4 shows the points obtained on die 1. Die 2 displays its points within the range H2:J4.

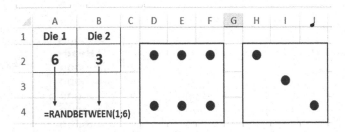

FIGURE 8.21   Drawing dice.

=IF(A2>1;CHAR(149);"")   =IF(A2=6;CHAR(149);"")   =IF(A2>3;CHAR(149);" ")

                            =IF(OR(A2=1;A2=3;A2=5);CHAR(149);" ")

=IF(A2>3;CHAR(149);"")   =IF(A2=6;CHAR(149);"")   =IF(A2>1;CHAR(149);"")

FIGURE 8.22   Generator expressions for the dice. Range D2:J4.

Figure 8.22 presents the expressions that arrange the points of die 1. In each cell, the contents of cell A2 are evaluated. To display the points in die 2, the same expressions are repeated in the range of cells H2:J4, although the expression uses the result of cell B2. By repeatedly pressing the function key F9, the top face of the pair of dice is dynamically simulated.

Example 15: Guess the Key and Win Contest

A contestant receives 10 keys, one of which is a key to open a door and win a motorcycle. The contestant has 3 chances with repositioning, so that each time he must randomly choose one of the 10 keys and test if he can open the door. If the selected key is not the correct one, it is returned and mixed in the group of keys to then make the next attempt of three chances. (A) For a confidence level of 99% and an error of 1% determine the length of simulation required to estimate the probability that the participant wins the contest. (B) Simulate and estimate the probability of winning the contest. (C) Theoretically, what is the probability that the participant wins the contest?

Solution

A. Determination of the sample size or simulation length:

The parameter to be estimated corresponds to a probability. The expression:

$$n \geq \frac{p * (1 - p) * Z_{\frac{\alpha}{2}}^2}{\in^2} = \frac{Z_{\alpha/2}^2}{4 * \in^2}$$

Assumes p = 0.5. That is, equal probability is assigned to the Win or Lose events. If desired, a pilot sample allows estimating the p-value. However, with 0.50 the product p*(1 − p) takes its maximum value 0.25 = 1/4.

The confidence level is $1 - \alpha = 0.99$ from which $\alpha/2 = 0.005$ is obtained. The error $\epsilon = 0.01$. Consequently,

$$n \geq \frac{NORM.S.INV(0.005)^\wedge 2}{4 * 0.014 * 0.01} = \frac{6.634897}{0.0004} = \frac{6.634897}{0.0004} = 16,588 \text{ participants}$$

Therefore, the length of the simulation must be at least 16,588 participants.

B. For the calculation of the probability, Figure 8.23 shows the calculated values. Cell B1 receives the number of keys given to the participant. After simulating the participation of 16,588 participants, cell J2 gives the estimated probability of winning: 0.2761.

C. The calculation of the theoretical probability of a participant winning is calculated according to the number of keys he receives by the following expression:

$$= 1/B1 + (1/B1) * ((B1 - 1)/B1) + ((B1 - 1)/B1)^2 * (1/B1)$$

In this case, B1 = 10 keys. That is, the probability that a participant succeeds in opening on the first, second, or third attempt is exactly:

$$1/10 + (9/10) * (1/10) + (9/10)^2 * (1/10) = 0.271$$

This is shown in cell F1. According to the conditions of the statement, the simulation should estimate that probability with a margin of error of 1/100th out of 99% of the time.

The alternatives available to get a computer to execute a certain procedure are numerous. Here is a different approach to simulate the above exercise. The previous simulation focuses on the probability of the participant selecting the correct key. In this alternative case for each participant, the initial number representing the key that achieves the opening indicated in the cells of column B is simulated. Then, in the cells of column C, the integer values corresponding to the key that the participant chooses at each opportunity are generated; so that if any value coincides with the opener key, it is because it corresponds to the key that will open the door (see Figure 8.24).

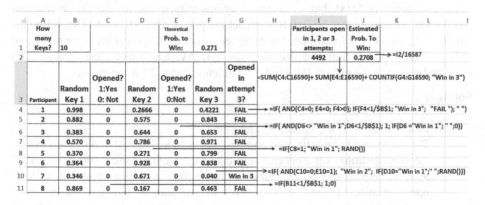

FIGURE 8.23 Guess the key and win a motorcycle.

	A	B	C	D	E	F	G
1	Participant	Key which Open	Open in 1, 2 or 3 Attempts?		Estimated Prob(Open in 1, 2 or 3 Attempts)		
2	1	2	Y		0.000180865		
3	2	4	N				
4	3	5	Y		=COUNTIF(C2:C16588;"Y")/16587		
5	4	6	Y				
6	5	7	N				
7	6	3	N		=IF(RANDBETWEEN(1;10)=B2;"Y";IF(RANDBETWEEN		
8	7	1	N		(1;10)=B2;"Y";IF(RANDBETWEEN(1;10)=B2;"Y";"N"))		
9	8	9	N				
10	9	8	N		=RANDBETWEEN(1;10)		

FIGURE 8.24   Simulation alternative for the contest.

A more realistic version of the quiz could consist of assuming that the participant has a good memory and giving him at once a bunch of 10 keys, so that when he fails, he remembers and discards the failed key. Thus, the theoretical probability would be calculated according to:

$$\frac{\text{Number of attempts allowed}}{\text{Number of keys}} = 3/10$$

For a sample size of 16,588 participants, simulate to estimate the expected probability that a participant with good memory wins the motorcycle in 1, 2, or 3 attempts.

Example 16: Casino Slot Machine

A slot machine in a casino consists of three independent rotating cylinders. Each cylinder has four equally distributed figures: lemon, star, seven, and cherries. When the gambler presses a button, the three cylinders start spinning and after many spins they stop randomly, each one showing one of the four figures. If the player obtains three equal figures, he receives $500; but if the three figures correspond to sevens, he receives $1,000. Otherwise he loses his bet. In each bet the player risks $100. (A) Determine the length of the simulation to estimate the expected utility of the player for a confidence of 95% and error of $3. Simulate and estimate: (B) Average utility of the player. (C) Expected utility of the player. (D) Find a confidence interval for the utility.

Solution

A. Determination of the simulation length:
   The parameter to be estimated corresponds to an average. It is observed that the variable utility for a bet can only take 3 different values: −100, 400, and 900. Moreover, the utility shows a distribution different from that of a normal statistical distribution. Therefore, the sample size to simulate is obtained from the Chebyshev expression with confidence level $1 - \alpha = 0.95$,

where α = 0.05 and error ∈ = 3. The variance S^2 is estimated from a pilot sample by executing 1,000 bets.

$$n \geq \frac{S^2}{\alpha * \epsilon^2} = \frac{22,683}{0.05 * 3 * 3} = 50,407 \text{ games}$$

B. A cylinder has 4 figures each one with equal probability of appearing; that is, one of four which is equivalent to 0.25. A procedure consists of generating a random value to imitate the figure that appears in each cylinder and determining according to the criteria expressed in Figure 8.25.

Cell B9 (Figure 8.26) contains the random value generated to mimic the output of cylinder 1, whose result lemon appears in cell E9, obtained through the function =LOOKUP. This function searches the random value contained in cell B9, in the search uses cells $E$1 to $H$1.

C. The average utility is obtained through the function =AVERAGE applied on the 50,407 bets. The value −60.839 in cell H4 indicates that the player has an average loss of approximately $61 each time he presses the slot button. If 20 runs are executed it could average around $ −60.60. However, the game expected utility is a loss of $ −60.94, which could be analytically computed as in Figure 8.27:

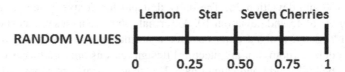

FIGURE 8.25 Procedure to imitate the spinning and stopping.

	A	B	C	D	E	F	G	H	I	J	K	L
1		Prize $			Lemon	Star	7	Cherry				
2	3 Sevens:	1000		0	0.25	0.50	0.75	1				
3	3 Equal Figures:	500										
4					E(Utility) $			-60.839	=AVERAGE(H9:H50415)			
5	Bet Amount $:	100			=IF(AND(E9=7;F9=7;G9=7);$B$2-$B$5;IF(AND(E9=F9;F9=G9);$B$3-$B$5;-$B$5))							
6												
7		Random Generation			Figure in Cylinder							
8	Game	R1	R2	R3	Cylinder 1	Cylinder 2	Cylinder 3	Utility				
9	1	0.19	0.20	0.60	Lemon	Lemon	7	-100	=LOOKUP(B9; $D$2:$H$2; $E$1:$H$1)			
10	2	0.29	0.14	0.72	Star	Lemon	7	-100				
11	3	0.36	0.21	0.11	Star	Lemon	Lemon	-100				
12	4	0.24	0.01	0.15	Lemon	Lemon	Lemon	400				

FIGURE 8.26 Sample simulation of the slot machine.

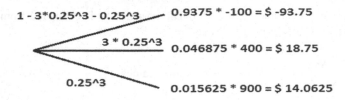

$1 - 3*0.25^3 - 0.25^3$ ⟶ $0.9375 * -100 = \$ -93.75$

$3 * 0.25^3$ $0.046875 * 400 = \$ 18.75$

0.25^3 $0.015625 * 900 = \$ 14.0625$

FIGURE 8.27   Theoretical expected game utility.

D. The function:

=CONFIDENCE.NORM(Alpha; deviation; size)

Allows to establish a confidence interval for the expected utility of the game:

=CONFIDENCE.NORM(0.05;STDEV.S(H9:H50415);50407) = 1.395

Thus, with 95% confidence the expected utility of the game should be between $ −62.23 and $ −59.44. [−60.839±1.395].

Example 17: Hotel Rooms Remodeling
The manager of a luxury hotel evaluates its remodeling. He has area for 200 normal rooms. However, he can incorporate deluxe rooms. A deluxe room is equivalent area of two standard rooms. A standard room rents for $990/night and a deluxe room for $1,800. The daily demand for standard rooms is random normally distributed with an average of 100 and deviation 25. The daily demand for deluxe rooms is also normal with an average of 45 and deviation 10. A normal room incurs a setup cost of $75; and a deluxe $140 daily. A fixed daily cost for a normal room is estimated at $550 and $1,000 for a deluxe room. For architectural design reasons only multiples of 10 rooms per type can be built, and hotel cannot have less than 50 standard rooms. Simulate and determine the optimum number of rooms to be provided.

Solution
The Solver program integrated into Excel offers a direct alternative to solve these types of optimization problems. This plugin can be activated from the toolbar via FILE / Options / Plugins / Go. Once the plugin is activated, it appears on the Data tab, in the Analysis group. Solver requires declaring a cell that contains the target value to optimize. In addition, it requires indicating all the cells that contain the variables that can be varied to achieve optimization. Finally, the restrictions to be met by the variables will be established.

The provision of building at least 50 normal rooms and only a multiple of 10 rooms reduces the building options to the following combinations (Table 8.4):

TABLE 8.4 Building options

Standard	200	180	160	140	120	100	80	60
Luxurious	0	10	20	30	40	50	60	70

Each of the eight alternatives will be simulated for 50,000 days. For example, alternative (180;10), indicates that during each day at the beginning, a fixed number of 180 normal rooms and 10 luxury rooms are available. During each day the demand for rooms by type is generated and revenue and expenditure calculations are made to determine the profit.

After simulating the 50,000 days the average utility value corresponding to the alternative (180;10) is obtained. The same procedure is applied to the remaining 7 combinations to finally obtain the one that gives the maximum average value. Figure 8.28 contains a partial sample of results when simulating the option (180;10).

In Figure 8.28, cells I3 and J3 store the values of the decision to build 180 standard and 10 deluxe rooms. Both cells contain the values of the decision variables. If you want to use the Excel Solver add-in, the range of the variables would be I3:J3. Cell L1 would be declared as the target.

For a run of length 50,000 days, cell **L1** contains an estimate of average profit of $ −3,318.42. By repeating the simulation procedure for the remaining 7 policies, we are able to select the best building alternative.

To evaluate each possible option, build an Excel table. See Figure 8.29. Once values to be evaluated are placed in ranges N2:N9 and O1:V1, select table range N1:V9 and apply from the toolbar.

	A	B	C	D	E	F	G	H	I	J	K	L
1	Daily Prize x Room:		Demand/day:		Daily Cleaning Cost:		Fixed Cost:		Option Simulated (N;L):		Average Profit $/day:	-24,048.28
2	Normal	Deluxe	Normal	Deluxe	Normal	Deluxe	Normal	Deluxe	Normal	Deluxe		
3	990	1800	100	45	75	140	550	1000	180	10		=AVERAGE(L10.L50009)
4			25	10							=F10+G10-H10-I10-J10-K10-$I$3*$G$3-$J$3*$H$3	
5	=INT(NORM.INV(RAND();$C$3;$C$4))						=E10*$B$3					
6					=IF(C10>$J$3; $J$3; C10)		=D10*$E$3		=IF(B10>$I$3; (B10 -$I$3)*($A$3 -$G$3 -$E$3); 0)			
7												
8			Room Demand		Rooms Used		Income		Room Set Up Cost		Oportunity Cost	
9	Day	Normal	Deluxe	Normal	Deluxe	Normal	Deluxe	Normal	Deluxe	Normal	Deluxe	Profit/day
10	1	125	50	125	10	123,750.00	18,000.00	9,375.00	1,400.00	-	26,400.00	-4,425.00
11	2	103	32	103	10	101,970.00	18,000.00	7,725.00	1,400.00	-	14,520.00	-12,675.00
12	3	77	40	77	10	76,230.00	18,000.00	5,775.00	1,400.00	-	19,800.00	-41,745.00

FIGURE 8.28 Simulation of option (180;10).

	M	N	O	P	Q	R	S	T	U	V
1	=L1	-24,048.28	200	180	160	140	120	100	80	60
2		0	-48,043.23	-37,347.31	-26,519.31	-15,990.39	-8,030.11	-5,719.42	-11,340.01	-22,637.04
3		10	-35,123.29	-24,043.28	-13,199.01	-2,823.63	5,225.06	7,410.77	1,944.89	-9,462.98
4		20	-21,914.06	-10,948.83	-137.37	10,149.14	18,450.17	20,688.76	15,067.94	3,557.93
5		30	-9,450.03	1,512.00	12,343.30	22,783.67	30,661.44	33,094.64	27,617.27	16,144.24
6		40	-314.38	10,559.90	21,595.23	31,916.92	39,780.17	41,991.56	36,555.67	25,141.34
7		50	672.32	11,684.39	22,705.19	32,934.93	40,881.79	43,254.03	37,662.67	26,314.73
8		60	-5,734.13	5,133.12	16,282.92	26,464.14	34,678.51	36,732.88	31,275.44	20,022.71
9		70	-15,250.30	-3,907.11	6,769.24	17,061.30	25,435.98	27,589.46	21,792.96	10,587.68

FIGURE 8.29 Two-entry table for hotel utilities.

DATA/What–If Analysis/Data Table … /Row input cell **I3**/Column input cell **J3**/OK

The cells with possible values are highlighted in bold type. It can be seen in Figure 8.29, cell T7. The decision with the highest average profit is to build 100 standard rooms and 50 deluxe rooms, with an expected profit value of $43,254.03 per day. This value is a random result from a size one random sample. To obtain best approximation for the utility it is recommended to run and average at least 20 repetitions.

Example 18: Bicycles Production and Quality

In a warehouse there are six good bicycles and four defective ones. A quality inspector randomly selects three bicycles. Simulate 50,000 inspections to estimate the probability that exactly two of the three bicycles are defective.

Solution

One of the different ways of approaching this problem using the Excel sheet is to assume that the selection of bicycles is done one by one. Figure 8.30 shows the report of 5 out of 50,000 inspections. In the first inspection, a D appears in cell C8, indicating that the first bicycle chosen was defective because the random value of cell B8 is less than 0.4, the probability of obtaining a defective bicycle in the first selection.

To simulate the selection of bicycle 2, it is required to determine whether the first selection was defective or good, which conditions the search for the random number of cell D8. If the first extraction is defective, then the probability of selecting a defective one in the second one would be 3/9, and if it was good 4/9, values that appear in cells C2 and C3, respectively. This implementation is declared in cell E8 by the following Excel expression:

=IF(C8="D";SEARCH(D8;$B$2:$D$2;$C$1:$D$1);SEARCH(D8;$B$3:$D$3;
$C$1:$D$1))

Before extracting the third bicycle, the number of accumulated defective bicycles is obtained (Cells F8:F50007). This is for the purpose of deciding the search probabilities.

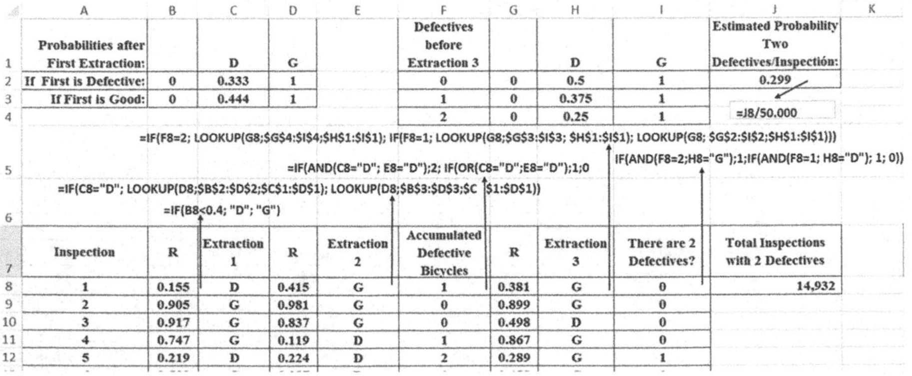

	A	B	C	D	E	F	G	H	I	J	K
1	Probabilities after First Extraction:		D	G		Defectives before Extraction 3		D	G	Estimated Probability Two Defectives/Inspectión:	
2	If First is Defective:	0	0.333	1		0	0	0.5	1	0.299	
3	If First is Good:	0	0.444	1		1	0	0.375	1		
4						2	0	0.25	1	=J8/50.000	
5											
6											
7	Inspection	R	Extraction 1	R	Extraction 2	Accumulated Defective Bicycles	R	Extraction 3	There are 2 Defectives?	Total Inspections with 2 Defectives	
8	1	0.155	D	0.415	G	1	0.381	G	0	14,932	
9	2	0.905	G	0.981	G	0	0.899	G	0		
10	3	0.917	G	0.837	G	0	0.498	D	0		
11	4	0.747	G	0.119	D	1	0.867	G	0		
12	5	0.219	D	0.224	D	2	0.289	G	1		

=IF(F8=2; LOOKUP(G8;$G$4:$I$4;$H$1:$I$1); IF(F8=1; LOOKUP(G8;$G$3:$I$3; $H$1:$I$1); LOOKUP(G8; $G$2:$I$2;$H$1:$I$1)))

=IF(AND(C8="D"; E8="D");2; IF(OR(C8="D";E8="D");1;0)

IF(AND(F8=2;H8="G");1;IF(AND(F8=1; H8="D"); 1; 0))

=IF(C8="D"; LOOKUP(D8;$B$2:$D$2;$C$1:$D$1); LOOKUP(D8;$B$3:$D$3;$C $1:$D$1))

=IF(B8<0.4; "D"; "G")

FIGURE 8.30 Inspection of three bicycles.

If 2 defective ones have been accumulated, the probability of extracting the third defective bicycle is 2 out of 8, or 0.25. But if the first 2 were good, then 4 defective ones remain out of the 8 bicycles, which gives a probability of 0.5 to obtain a defective bicycle or 4/8. These conditional decisions are expressed in column H, from cell H8.

Finally, column I from cell I8, verifies the favorable event of interest. That is, if the inspection yields exactly two defective bicycles, for which, if the answer is yes, a 1 is recorded; otherwise a zero is displayed. Cell J8 adds up all the 1s in cells I8 to I50007, a value that is used in cell J2. This value divided by the 50,000 simulated inspections gives an estimate of the probability of extracting exactly two defective bicycles in any inspection. In this run, 0.299 was obtained. If the simulation is repeated for 19 more runs, the average of the 15 values is obtained for an estimate closer to the theoretical value 0.30.

Example 19: Cashier Service
At a checkout service, customers arrive at the Poisson rate of 4 customers per hour. The checkout counter is attended by a cashier who serves customers in an exponential average time of 12 minutes/customer. Simulate the service to determine: (A) The average waiting time of a customer. (B) The average number of customers waiting or expected length of the queue. (C) Occupancy factor of the system.

Solution
Figure 8.31 shows that for the first customer his arrival time appears in cell D5 and takes the value =C5. The waiting time of the first customer is placed in cell E5 with a value of null. Cell J5 takes as time in the system the content of G5, since it is the time of attention of the first customer.

Figure 8.31 presents partial simulation of arrival and service of only five customers. After extending the simulation to 100,000 customers the required responses are obtained.

A. The waiting time Wq is obtained by averaging the waiting values that appear in column E by means of =AVERAGE(E5:E100004). This value can be approximated to 48 minutes/customer.

B. The average queue length Lq is estimated from the quotient of the sum of all waiting times =SUM(E5:E100004), divided by the Last Customer Start Time value G100004. This value will approximate 3.2 customers.

Customer	Random (Arrivals) =RAND()	Time between Arrivals =-$A$2*LN(B5)	Time of Arrival =C6+D5	Wait Time =G6-D6	Service Time =-$B$2*LN(RAND())	Begin Service at =MAX(H6;D7)	Exit Time =G7+F7	System Time =H6-D6
1	0.425	12.8	12.8	0.0	12.6	12.8	25.4	12.6
2	0.293	18.4	31.3	0.0	10.5	31.3	41.8	10.5
3	0.045	46.4	77.6	0.0	13.6	77.6	91.2	13.6
4	0.957	0.7	78.3	12.9	12.7	91.2	104.0	25.7
5	0.467	11.4	89.7	14.3	17.2	104.0	121.2	31.5

FIGURE 8.31 Customer service and waiting with a server.

C. The occupancy factor ρ is achieved from the quotient of the sum of all service times =SUM(F5:F100004) divided by the last customer's departure time cell H100004; this value will be around 0.8.

It should be repeated about 20 times more to average and estimate those statistics.

Example 20: Product Inventory Policy (Q,s)

The daily demand for a product follows a normal random variable with average of 45 units and deviation of 10. The ordering policy of the stockist is as follows: At the end of each day, the stock is checked and if the level is 50 units or less a request for 100 units is issued to the supplier. As long as there is a request pending to arrive from the supplier, another order is not issued. Although sometimes the order arrives and is available the next day, the number of days it takes to arrive is random exponential average 2 days/order.

Inventory starts with a stock of 100 units. The purchase cost per unit is $50. The fixed cost of issuing an order to the supplier is $1,000. If there are not enough units in stock, the demand is partially satisfied with the quantity in stock, although for each unit not supplied an opportunity cost of $10 is incurred. The cost of keeping a unit in inventory is $5/day. The unit selling price is $75. Simulate an appropriate length and estimate: (A) Expected daily utility of the current policy. (B) Probability that the current policy does not allow to meet the total demand on any given day. (C) Using an Excel table determine the values of quantity to order Q and replenishment level R which optimize the daily utility of the inventory.

Solution

By taking a pilot sample of length 1,000 days we can see that the variable daily utility has mean, mode, and median very different and should not be assume to follow a normal distribution. So, the Chebyshev theorem will be used to define how many runs to execute. The sample variance of daily utility, calculated from the pilot sample of 1,000 values yields a value of 255,869 days. In order to have 95% confidence and an estimate of the average daily utility with error less than $5, the length of the simulation is obtained applying the Chebyshev relation being necessary to simulate the process for at least 39,318 days.

$$\text{length } n \geq \frac{S^2}{\propto * \, \in^2} = \frac{255,869}{0,05 * 25}$$

$$n \geq 39,318 \text{ días}$$

Figures 8.32 and 8.33 only show a simulation of length 6 days. However, it will be simulated 50,000 days.

Solving this problem requires evaluating various policies. Figure 8.31 shows the input values to be simulated. It starts with the policy (Q;R) = (100;50) corresponding to ordering 100 units when the stock is 50 units or less. The range A2:H2 contains the input data. A simulation length of 50,000 days is assumed. Each policy will be evaluated for 50,000 days to determine its net utility.

	A	B	C	D	E	F	G	H
1	Quantity To Order Q	Reorder Level R	Delay to Receive	Average Demand	Normal Deviation	Initial Stock	Unit Cost	Order Cost
2	100	50	2	45	10	100	50	1,000.00
3								
4	Day	Pending Order? Y/N	Initial Stock	Daily Demand	Final Stock	Make an Order? Y/N	Delay to Arrive	Day for Available Order
5	1	N	100	46	54	N	--	--
6	2	Y	54	51	3	Y	0	3
7	3	N	103	51	52	N	--	3
8	4	N	52	53	0	Y	2	7
9	5	Y	0	36	0	N	--	7

FIGURE 8.32 Simulation of an (Q,R) inventory policy.

	I	J	K	L	M	N	O	P
1	Shortage Cost	Holding Cost	Selling Price		Average Daily Utility	Shortage Probability		
2	10.00	5.00	75.00		150.12	0.427		
3								
4	Sold Units	Shortage Units	Sale Cost	Holding Cost	Shortage Cost	Order Cost	Sales Income	Day Utility
5	46	0	2300	270	0	0	3450	880
6	51	0	2550	15	0	1000	3825	260
7	51	0	2550	260	0	0	3825	1015
8	52	1	2600	0	10	1000	3900	290
9	0	36	0	0	360	0	0	-360

FIGURE 8.33 Continuation sheet of the inventory policy.

B5: N   'Enter N in cell B5 there are no pending orders
C5: =F2N   'allocate beginning inventory from F2
D5: = INT($D$2+NORMSINV(RAND())*$E$2)   'calculates daily demand integer' value
E5: =IF(D5>C5; 0; C5-D5)   'determines if the stock is sufficient and updates it
F5: =IF(E5<=$B$2; "Y"; "N")   'check if you have to order from the supplier

G5: =IF(F5="Y";INT(-$C$2*LN(RAND())));"--")   'determine the entire delivery time

H5: =IF(G5="--"; "--"; G5+1+A5)   'calculates the next day of admission of the supplier

I5: =IF(C5>=D5; D5; C5)   'record units sold

J5: =IF(C5<D5; D5-C5; 0)   'record missing units

K5: =$G$2*I5N   'cost of goods sold

L5: =$J$2*E5N   'holding cost}

M5: =$I$2*J5N   'cost of shortage

N5: = IF(F5="Y";$H$2;0)   'if the supplier is asked, to add cost per order

O5: =$K$2*I5N   'calculate sales revenue

P5: =O5-K5-L5-L5-M5-N5   'calculate daily profit

'The cells in row 6 to be copied to extend the simulation to 50,000 days:

A6: =A5+1

B6: =IF(H5>A6; "Y"; "N")

C6: =IF(A6=H5;E5+$A$2;E5)

D6: = INT($D$2+NORMSINV(RAND())*$E$2)

E6: =IF(D6>C6;0;C6-D6)

F6: = IF(AND(E6<=$B$2; H5<=A6); "Y";IF(H5="--"; "Y"; "N")) 'If existence is less 'than the reorder point R, and also the day for supplier arrival is less than or equal to the 'current day, place Y to order. Otherwise if H5 contains -- set Y

G6: = IF(F6="Y";INT(-$C$2*LN(RAND())));"--")

H6: =IF(G6="--"; H5; G6+1+A6)

I6: =IF(C6>=D6; D6; C6)

J6: =IF(C6<D6; D6-C6; 0)

K6: =$G$2*I6

L6: =$J$2*E6

M6: =$I$2*J6

N6: = IF(F6="Y";$H$2;0)

O6: =$K$2*I6

P6: =O6-K6-L6-L6-M6-N6

The content of the formulas arranged in each cell of rows 5 and 6 is as follows:

A. To obtain the expected utility, the utilities corresponding to the 50,000 days of simulation are averaged using the formula in Figure 8.33, cell M2:= AVERAGE(P5:P50004).

B. The probability of shortage with the current policy is estimated in cell N2, see Figure 8.33. For this, it is required to check column J in order to count the days in which units are missing to fully meet the demand and divide by the total number of days according to the following expression: =COUNT.SI(J5:J50004;">0")/50000.

C. To determine the optimal policy, the double entry table shown in Figure 8.34 is constructed. The quantity to order Q is added in cells of row 2 for values

	R	S	T	U	V	W	X	Y	Z	AA	AB	AC
1	Reorder Point	=M2				Quantity to Order to Supplier from 110 to 210						
2	259.64	110	120	130	140	150	160	170	**180**	190	200	210
3	45	170.87	203.10	222.93	232.27	242.12	251.94	252.63	255.87	258.54	253.45	248.82
4	50	177.48	206.14	220.67	233.39	245.42	251.61	258.69	257.36	257.14	256.92	246.21
5	**55**	175.56	213.31	227.11	239.78	246.41	254.08	257.26	**259.88**	253.83	254.44	250.32
6	60	180.55	208.54	229.08	239.86	247.11	252.23	249.73	254.68	255.51	250.54	246.67
7	65	187.68	209.60	228.20	241.21	250.18	250.20	257.45	254.42	253.44	251.07	243.08
8	70	189.51	208.04	229.85	245.77	244.51	246.57	253.74	251.53	248.06	240.80	237.24
9	75	188.67	209.21	228.30	239.31	244.30	248.21	246.72	251.75	246.26	241.53	235.69
10	80	183.86	209.81	224.00	234.40	244.46	242.64	245.32	243.94	240.27	235.66	226.57

FIGURE 8.34 Double entry table: (Q, r).

from 110 to 210 units, 10 by 10; and the replenishment level in cells of column R, from 45 to 80 units, 5 by 5. Cell R2 contains the value of cell M2. The table requires after selecting the range R2:AC10 to choose from the toolbar the option:

Data / What-if Analysis / Data Table ... / Row input cell: **A2** / Column input cell: **B2** / OK

The table in Figure 8.34 presents as maximum value of average profit $259.88, corresponding to the policy (180; 55). However, since this amount corresponds to a random result among countless possible results, it is necessary to repeat the experiment about 20 times in order to average the 20 values obtained by each policy. Finally, compare and evaluate the average utility of each alternative to determine the optimal policy.

Example 21: True or False Questionnaire
A questionnaire consists of 10 questions the answers to which are true or false. A person with no knowledge of the subject decides to choose his answer according to the outcome of a coin toss. Simulate and estimate the probability that a test taker gets at least 5 questions correct.

Solution
The procedure starts by setting in cells B2 to K2, Figure 8.35, 10 correct answers to the test. To imitate the answers of an irresponsible test taker who answers according to the result of a coin toss a uniform random value is selected. If the chance value is less than 0.5, it is assumed that the respondent answered True (T). Otherwise, he/she answered False (F).

Cells B6 to K6 (Figure 8.35) contain the formulas that mimic the answers given by the first person evaluated. That is, cell **B6** contains =**IF(RAND()<0.5; "T"; "F")**. Similarly, to simulate the responses of 50,000 test takers, cell **B6** is copied into the range **B7:K50005**.

	B	C	D	E	F	G	H	I	J	K	L	M	N	O	P	Q	R	S	T	U	V	W
A																						
1 Test Question Number	1	2	3	4	5	6	7	8	9	10												Probability at least 5 Rigth
2 Correct Answer	T	F	F	T	F	T	T	F	T	F												0.6216
3																						
4			Respondent Answers to questions 1 a 10									Evaluation				(1--> Correct 0 --> Failure)						Respondent Rigth Answers
5 Respondent	1	2	3	4	5	6	7	8	9	10		1	2	3	4	5	6	7	8	9	10	
6 1	F	T	F	T	T	T	T	T	F	F		0	0	1	1	0	1	1	0	0	1	5
7 2	T	F	F	F	F	T	T	T	T	F		1	1	1	0	1	1	1	0	1	1	8
8 3	F	T	F	F	F	T	F	F	F	F		0	0	1	0	1	1	0	1	0	1	5
9 4	F	F	T	T	T	F	T	T	T	F		0	1	0	1	0	0	1	0	1	1	5
10 5	T	F	T	T	T	T	F	T	T	F		1	1	0	1	0	1	0	0	1	1	6

FIGURE 8.35   Simulation of responses in a V/F test.

Columns M to V compare the answers of each test taker with the exact answers to the test. A value of 1 is assigned to a right answer and 0 otherwise. The displayed formulas have the following expressions:

M6: =IF(B6=$B$2; 1; 0)	N6: =IF(C6=$C$2; 1; 0)
O6: =IF(D6=$D$2; 1; 0)	P6: =IF(E6=$E$2; 1; 0)
Q6: =IF(F6=$F$2; 1; 0)	R6: =IF(G6=$G$2; 1; 0)
S6: =IF(H6=$H$2; 1; 0)	T6: =IF(I6=$I$2; 1; 0)
U6: =IF(J6=$J$2; 1; 0)	V6: =IF(K6=$K$2; 1; 0)

Determining the number of correct answers requires summing the values 1 in the row of the evaluated. For this, cell W6: =SUM(M6:V6).

Estimating the probability that a test taker gets 5 or more questions right requires establishing whether the total number of correct answers is 5 or more. Cell **W2** contains the conditional expression: **=COUNTIF(W6:W23;">=5")/50000**, which counts the number of cells in column W, whose sum equals or exceeds 5; and then divides by 50,000 test takers to give the probability sought. In this case, the probability that an individual who chooses at chance obtains 5 or more right answers is 0.6216. Again, we insist on recognizing that the result corresponds to a random value unique among an infinite number of possible answers. For an acceptable estimate a single result is insufficient. Average at least 20.

Example 22: UN: Human Development Index
The United Nations through the United Nations Development Program (UNDP) calculates the Human Development Index annually. The HDI is based on three indicators: longevity (healthy and prolonged life), educational level (possibility of acquiring knowledge), and standard of living (worthy). Between 1970 and 2000, the HDI of the different countries was collected and classified into three states: high if HDI >= 0.80; medium if $0.5 \le$ HDI ≤ 0.799, and low if HDI < 0.5. For 5-year periods the following transition probability matrix was obtained:

n/n + 1	L	M	H
L	0.885	0.115	0
M	0.084	0.896	0.02
H	0	0.178	0.822

The process corresponds to a three-state Markov Chain and the table shows conditional transition probabilities between the possible states. For example, if a country in the current 5-year period **n** has a high HDI, the probability that in the next 5-year period **n + 1** will evolve to a medium HDI is 0.115. Simulate this Markov Chain process and estimate the expected long-term behavior of HDI.

Solution
From 5 years to 5 years the HDI of a country is observed in one or another of the three possible states. Once the cumulative probabilities for each of the states have been arranged, the model evolves over time in 5-year time lapses. Cell B6 (Figure 8.36), contains the starting HDI, which will be plotted in cell B8 as the HDI at the beginning of the 5 years. To determine the possible evolution at the end of the 5-year period, a random value is generated in cells of column C. This value originates a conditional search to the previous state, which appears in cell D8 arranged to obtain the HDI at the end of the 5-year period according to the following formula:

=IF(B8="A";SEARCH(C8;$B$2:$D$2;$C$1:$D$1);IF(B8="M";SEARCH(C8; $B$3:$E$3; $C$1:$E$1); SEARCH(C8;$C$4:$E$4;$D$1:$E$1)))))

	A	B	C	D	E	F	G	H	I
1			H	M	L				
2	H	0	0.885	1	1				
3	M	0	0.084	0.98	1				
4	L	0	0	0.178	1				
5									
6	Country Initial HDI:	M			Long-term state transition Expected for any country:				
7	Five_Year	HDI at five-year Begin	Random Value RAND()	HDI Five_Year End	High HDI	Medium HDI	Low HDI		
8	1	M	0.38192	M	39.3%	54.5%	6.2%		
9	2	M	0.13484	M					
10	3	M	0.67823	M	=COUNTIF(D8:D50007;"L")/50000				
11	4	M	0.00743	H					
12	5	H	0.77773	L					
13	6	L	0.63245	L	=IF(B8="H";LOOKUP(C8;$B$2:$D$2;$C$1:$D$1);				
14	7	L	0.10145	M	IF(B8="M";LOOKUP(C8;$B$3:$E$3;$C$1:$E$1);				
15	8	M	0.16059	M	LOOKUP(C8;$C$4:$E$4;$D$1:$E$1)))				

FIGURE 8.36 Human Development Index.

Note that the state of the end of one 5-year period corresponds to the initial state of the next. This is achieved, for example, by bringing into cell B9 the expression =D8 which is the state of the previous 5-year period.

After 50,000 five-year periods, estimating the long-term state transition probabilities only requires checking column D to count the number of times the country closed the 5-year period in each of the three possible states, and dividing by the number of 5-year periods considered. In this case 50,000. The expressions:

E8: =COUNTIF(D8:D50007;"H")/50000
F8: =COUNTIF(D8:D50007;"M")/50000
G8: =COUNTIF(D8:D50007;"L")/50000

Give estimates of the stable or long-term probabilities. In this case, if the behavior is maintained, it is expected that 39.3% of the countries will reach high HDI level, 54.5% medium level, and 6.2% low; values coming from the count and division arranged in cells E8, F8, and G8.

Example 23: Evolution of an HCM Insurance Fund
A transnational banking company manages the Hospitalization, Surgery, and Maternity (HSM) insurance system of its employees in a common fund, which starts with $500,000. We want to simulate the evolution of the fund. The HSM expenses are covered by fixed contributions from the employees. However, if the contribution does not cover all the claims for the month, the missing amount is paid by the transnational company.

At the beginning the company has 5,000 employees. However, from one month to the next the workforce may increase (I), decrease (D), or remain the same (S) by a percentage that varies according to a uniform distribution between 1% and 2%. The probability that from one month to the next the labor force will increase is 0.49; that it will decrease is also 0.49 and 0.02 that it will remain the same.

The average cost of an HSM event per employee is random normal with an average of $30,000 and a deviation of $5,000. The number of claims in any month is a normal random variable with average of 100 and variance of 64. Each employee contributes monthly to the collective fund a fixed amount of $500 used to cover HSM expenses, with the company paying any difference. (A) Simulate the evolution of the fund during 50,000 months. Determine: (B) Average monthly amount that the transnational expects to disburse for HSM claims. (C) Probability of having between 92 and 108 monthly claims.

Solution
The input data are shown in the range of cells A2:G2 in Figure 8.37. Cell A2 starts month 1 with 5,000 employees. The contents of formulas in each cell are shown below:

B5: =A2 takes the number of employees declared for the start of the simulation.

C5: =C2 corresponds to the initial start-up contribution to the HSM fund.

D5: =INT(NORM.INV(RAND();100;8)) this expression simulates the number of claims occurred in the month by generating a normal random value of average 100 and deviation 8. It uses the function =INT() to show the random claims value.

	B	C	D	E	F	G	H	I	J	K	L
1	Contribution per employee	Initial Insurance Fund:	Labor force trend Inc/Dec/Sam	I	D	S				Paid Average $/month	Probability 92<Claims<108
2	500	500,000.00	0	0.49	0.98	1				2,619,972	0.709
3			State Probability:	0.49	0.49	0.02					
4	Employees at the month begining	Availability $ at month begining	Number of Claims/month	Amount to pay for claims	Employee Input	Amount to be paid by the company	Fund Available at the end of month	Labor Trend	% Temporal variation	Employees at end of month	Claims between 92 y 108? Yes=1 Not=0
5	5,000	500,000	105	3,236,625	2,500,000	236,625	-	I	1.41%	5,071	1
6	5,071	-	99	3,193,542	2,535,306	658,236	-	I	1.01%	5,122	1
7	5,122	-	100	3,942,900	2,560,799	1,382,101	-	I	1.35%	5,191	1
8	5,191	-	89	2,581,089	2,595,307	-	14,218	D	1.17%	5,130	0
9	5,130	14,218	94	2,227,236	2,565,059	-	352,041	I	1.44%	5,204	1
10	5,204	352,041	92	3,189,180	2,602,104	235,036	-	I	1.49%	5,282	1
11	5,282	-	107	3,536,136	2,640,770	895,366	-	I	1.87%	5,380	1
12	5,380	-	108	2,622,672	2,690,105	-	67,433	I	1.02%	5,435	1
13	5,435	67,433	99	3,018,015	2,717,582	233,000	-	D	2.00%	5,327	1
14	5,327	-	103	2,531,946	2,663,289	-	131,343	D	1.54%	5,245	1
15	5,245	131,343	101	3,458,341	2,622,252	704,745	-	I	1.24%	5,310	1

FIGURE 8.37 Evolution of the insurance fund.

E5: (INT(NORM.INV(RAND(); 30000; 5000)))*D5 mimics the expected cost of a claim, by generating a normal random value of average 30,000 and deviation 5,000. For simplicity, the integer part of the random value is taken. This result is then multiplied by the number of claims for the month. It is not necessary to individualize the amount for each claim.

F5: =$B$2*B5 obtains the monthly amount contributed by the employees. Multiplies the individual contribution by the monthly number of employees.

G5: =IF(C5+F5<E5; E5-C5-F5; 0) adds the fund balance with the employees' contribution and compares it against the amount to be paid for all claims. This determines the amount to be paid by the company.

H5: =IF(C5+F5>E5; C5+F5-E5; 0) determines the balance available at the end of the month after the claims have been covered.

I5: =LOOKUP(RAND();$D$2:$G$2;$E$1:$G$1) generates a uniform random value to perform a search in cells D2:G2 and its result of increase, decrease, or equal. This simulates the trend of variation in the working population.

J5: =IF(I5 <> "S"; (1+RAND())/100;0) Evaluates the contents of cell I5 to obtain a uniform random value between 1 and 2, representing the percentage of variation of the working population.

K5: =IF(I5="I"; B5*(1+J5); IF(I5="D"; B5*(1-J5); B5)) to determine the number of employees at the end of the month, check the trend of the population and apply the corresponding percentage of variation.

L5: =IF(AND(D5>=92; D5<=108); 1; 0) this expression nests the logical function AND() in an IF() function. If the cell D5, which contains the number of claims for the month, is between 92 and 108, then 1 is registered. Otherwise 0 is placed. This value serves to show the characteristic of any population with normal distribution, which states that approximately 68.26% of its values fall between the average and one standard deviation.

That is, since the number of claims in any month is a normal random variable of mean 100 and variance 64, then it is expected that 68.26% of the months the number of claims will be between 92 (100 − 8) and 108 (100 + 8).

Figure 8.37 shows the answer to questions B) and C) for a simulation of 50,000 months. Cell K2 indicates that the company expects to make a monthly payment of $2,619,912. In relation to question C) on probability, cell L2 contains a value of 0.709. That is, according to this random result, 70.9% of the months, between 92 and 108 claims are expected. This approximates the theoretical expected value of 68.26%.

Example 24: Subprogram for Multiple Repetitions

In each simulation run, the complex interaction behavior between the random variables offers random responses with high variability. In order to estimate acceptably parameters such as the monthly amount expected to be paid by the company, multiple repetitions are required. For this purpose the following Excel subprogram or macro is incorporated:

```
Sub Repeat()
    n = Val(InputBox("How many iterations? ",, 20))
    Columns("M:M").Clear
    For i = 1 To n
        Cells(i + 1, "M") = Range("H1")
    Next i
End Sub
```

The procedure to follow to create the subprogram in Excel is shown in Table 8.5.

TABLE 8.5 Procedure to create a subprogram in Excel

1. If the **Developer** tab does not appear in the ribbon, it is necessary to activate it. To do so:
   a. Click on the **Office** button
   b. **File**
   c. **Options / Customize Ribbon** / Select **Developer** / OK
2. In the Options Ribbon active the option **Developer**
3. Select **Macros / Macro name** Repeat
4. Click **Create**
5. Type between **Sub Repeat** and **End Sub**:
   n = Val(InputBox("How many iterations? ",, 20))

   Columns("M:M").Clear

   For i = 1 To n

     Cells(i + 1, "M") = Range("H1")

   Next i
6. Select **Run / Run Sub-User Form F5**. The result of the execution is shown in the Excel sheet. If you look at the bar or ribbon at the bottom of the computer screen, Microsoft Excel and Microsoft Visual tabs appear. By clicking on one or the other, you can enter or exit both environments to run or view results.

In this example, the average monthly contribution to be expected by the company is estimated. The average contribution of the company in a run of 10,000 months is calculated in cell H7 by means of =AVERAGE(G5:G10004). In addition, 20 independent repetitions are desired, the results of which will be displayed in column I, specifically in the range M2:M21.

The instruction **n** = Val(InputBox("How many repetitions?",, 20)) asks the user from the keyboard to indicate the number of repetitions to be performed. This stores the answer in the variable **n**. The instruction **Columns("M:M").Clear** erases all the previous values of the column M. Then, the following instructions define a repetitive cycle from i = 1 to i = 20 runs:

```
For i = 1 To n
    Cells(i + 1, "M") = Range("H1")
Next i
```

This iterative cycle starting with variable i = 1, stores the value contained in cell H1 and places it in cell M2. M2 receives the average of the first 10,000 months. Then for i = 2 it takes the average of run 2 for another 10,000 months, which will be stored in H1 and displays it in cell M3. It repeats **n** times, 20 times in this case, filling column M from M2 to M21 with the 20 average values of 10,000 months each. Finally, the user can define a cell that obtains the average of the 20 averages stored in the range cells from M2 to M21. This value corresponds to a better approximation to the value of the parameter to be estimated.

It is possible to improve the automation of the experiment by incorporating programming control buttons provided by Excel. **Form controls** are dialog formats to prompt the user for data input. In the toolbar of the Excel spreadsheet, in the Scheduler ribbon, the **Insert** option appears. In its selection list the available Form Controls are displayed. The procedure to follow is as follows:

Developer / Insert / Button (Form control) /

The cursor turns into a small cross. Position and click the cross cursor in the free area of the worksheet. Select the name of the macro **Repeat** and **OK**.

By right clicking on the button you can name the button. Type the name that identifies the button: Simulate and click outside the button. Click the button every time you want to run the repetitions even for values of **n** less or greater than 20.

To save the Excel sheet and its macro from Excel click on the Office button. Then Save as Macro Enabled Excel workbook.

Example 25: Car Repair Shop

An auto repair shop repairs cars exclusively for a dealership. At the beginning of each day the dealer sends cars for repair and takes the cars already repaired. The number of cars arriving at the workshop has the following random behavior:

Cars Arriving	0	1	2	3	4
Probability	0.10	0.50	0.25	0.10	0.05

Due to space restrictions, the agreement with the concessionaire obliges not to allow more than four cars in the workshop, so any surplus will be delivered independently by the concessionaire to a neighboring workshop. The garage removes at least one car/day but never more than two cars/day. When there is more than one car the probability of repairing exactly one car is 0.45 and of repairing two cars is 0.55. If there is only one car, it will be repaired during the day. Simulate the operation of the workshop and estimate: (A) Percentage of the time the workshop is idle. (B) Average daily number of cars in the shop.

Solution
Figure 8.38 gives simulation results for a length of 50,000 days. The percentage of idle time in the workshop is estimated to be 0.92% of the time. That is, in a 365-day year the shop is expected to have no cars for approximately 14.5 days/year: $0.0092*365 = 3.4$ days/year. As for the expected number of units in the system, on average at any instant of visit to the shop 3.1 cars are expected to be found. Both estimates can be improved by averaging over several repetitions.

	A	B	C	D	E	F	G	H	I	J
1		Cars shipped by dealer daily:	0	1	2	3	4		Workshop idle percentage	Daily average of cars in the workshop
2		0	0.05	0.4	0.85	0.95	1		0.92%	3.122
3		%	5%	35%	45%	10%	5%			
4										
5	Day	Cars before new ones arrive	Random arrival	Cars brought by the dealer	Cars at the start	Cars in excess of 4	Cars to be repaired in the workshop	Repaired cars in the workshop	Cars at the end of the day	Day Idle? Y/N
6	1	0	0.934	3	3	0	3	2	1	N
7	2	1	0.373	1	2	0	2	2	0	N
8	3	0	0.856	3	3	0	3	1	2	N
9	4	2	0.075	1	3	0	3	2	1	N
10	5	1	0.967	4	5	1	4	2	2	N

FIGURE 8.38 Automobile repair shop.

The expressions stated in each cell of Figure 8.38 are as follows:

Cells C1 through G1, contain the possible values for the number of cars shipped by the dealership at the start of a day.
Cells B2 to G2 contain the cumulative probabilities for the number of cars.
I2: =COUNTIF(J6:J50005; "Y")/50000 counts the idle days in the garage and divides by the 50,000 simulation days to get the percentage of idle days in the workshop.

J2: =AVERAGE(G6:G50005) averages the number of cars in the workshop.
B6: =2 declares two cars to be repaired as initial value.
C6: =RAND() generates a uniform random value between 0 and 1 to determine the number of cars sent by the dealer.
D6: =LOOKUP(C6;$B$2:$G$2;$C$1:$G$1) searches and determines the number of cars coming in.
E6: =B6+D6 sums existing cars and new arrivals.
F6: =IF(E6>4; E6-4;0) determines the number of cars going to the neighboring garage.
G6: =E6–F6 determines the number of cars to be repaired in the shop.
H6: =IF(G6>1; IF(RAND()<0.45; 1; 2); IF(G6=1; 1; 0)) gets the number of cars repaired during the day.
I6: =G6–H6 determines the number of unrepaired cars remaining at the end of the day.
J6: =IF(G6=0; "Y"; "N") evaluates whether the number of cars to repair is 0, which indicates S as an idle day.
B7: =I6 fetches the number of cars left at the end of the previous day as the start value of the day.

The cell range C6:J6 must be copied over the range C7:J7. Subsequently, the range A7:J7 is copied to A50005:J50005. The parameters of interest will appear estimated for the length of 50,000 days.

Example 26: Auto Insurance and Claims
The records of an automobile insurance company indicate that the probability that an insured will suffer an automobile claim is 0.18. In the event of an accident, the amount payable to the client is a percentage of the insured value. Table 8.6 presents the percentage payable and its probability of occurrence.

TABLE 8.6 Insurance and claims data

% of the insured value to be indemnified:	25	50	75	100
Probability:	0.40	0.30	0.15	0.15

The insured value of any vehicle varies according to the type of policy contracted and is randomly distributed Normal with an average of $50,000 and a deviation of 10,000. The amount charged per policy is 13% of the insured price. (A) Determine the length of the simulation to estimate the expected profit per vehicle with 95% confidence and an error of $500. (B) Simulate and estimate the profit per car of the insurer. (C) Suppose you want to determine the percentage you should charge per insured to obtain an expected profit of at least $2500 per car.

Solution

To determine the simulation size a pilot run of 1,000 policyholders will be run. The parameter to be estimated, profit, corresponds to an average. It must be evaluated whether the 1,000 values of utility per insured resemble a normal distribution or not. Figure 8.39 presents the first five simulation rows of the pilot run.

After simulating the insurance policies for 1,000 policyholders, the Excel option **Data / Data Analysis / Descriptive Statistics** or **Histogram** can be used to evaluate the behavior of variable utility per policyholder, range G7:G1006.

Figure 8.40 contains the statistical values representative of the 1,000 utility values. They indicate that the utility variable does not follow a normal distribution where the mean, mode, and median values coincide. In addition, the degree of concentration of the values, its kurtosis of 8.27 is far from 0. The skewness coefficient −2.87 indicates that most of the values are more to the right, which is observed in the histogram. From which it is concluded that the utility values do not follow a normal distribution. Consequently, to estimate the simulation length the Chebyshev formula has to be used:

$$n \geq \frac{S^2}{\alpha * \in^2} = \frac{129,826,296.12}{0.05 * 500 * 500} = 10,387$$

Where S^2 corresponds to the sampling variance calculated with the pilot sample values. The simulation length will be carried out for 10,500 policyholders. After simulating the issuance of 10,500 policies, the average of column (Figure 8.39), range G7:G10506 is obtained. This value corresponds to the average profit per policyholder estimated at around $1,877.01.

To determine the percentage of the insured value that the insurer must charge to obtain an expected profit of at least $4,000 per car, it is sufficient in cell B4 to try different values of the percentage to be charged. A value of 17.25% allows this

	A	B	C	D	E	F	G	H
1	Accident Probability:	0.18	Percentage to Compensate:	25	50	75	100	
2	Insured Average:	50,000	Probability:	0.4	0.3	0.15	0.15	
3	Deviation	10,000		0	0.4	0.7	0.85	1
4	Percentage to Collect:	13						
5								
6	Insured	Amount Insured	Amount Charged	Had an Accident? Y/N	% to Compensate	Amount Paid for the Claim	Utility	Average Profit per Insured
7	1	45,202	5,876	Y	50	22,601	-16,725	1,877.01
8	2	45,715	5,943	N	0	-	5,943	
9	3	82,252	10,693	Y	75	61,689	-50,996	
10	4	62,541	8,130	N	0	-	8,130	
11	5	57,469	7,471	N	0	-	7,471	

FIGURE 8.39 Compensation for auto claims.

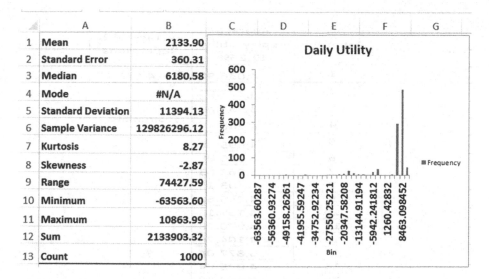

	A	B
1	Mean	2133.90
2	Standard Error	360.31
3	Median	6180.58
4	Mode	#N/A
5	Standard Deviation	11394.13
6	Sample Variance	129826296.12
7	Kurtosis	8.27
8	Skewness	-2.87
9	Range	74427.59
10	Minimum	-63563.60
11	Maximum	10863.99
12	Sum	2133903.32
13	Count	1000

FIGURE 8.40 Statistics and Excel histogram.

objective to be achieved. In order to accomplish a better estimate of the profit per insured, it is advisable to run at least 20 repetitions. For this purpose the following subprogram can be incorporated:

```
Sub Replik()
    n = Val(InputBox("How many iterations?",, 20))
    Columns("J:J").Clear
    For i = 1 To n
        Cells(i + 1, "J") = Range("H7")
    Next i
End Sub
```

Figure 8.41 shows in column J results of running several repetitions. Its cells from J2 to J51 were arranged to record the utility until 20 repetitions. In addition, the final calculation which averages the "averages", appears in cell K2. This gives a better estimate of the average profit per insurance policy of $4,031.47 if the collection rate were raised to 17.25% of the insured amount.

The Excel functions used to perform the simulation are in Table 8.7

Example 27: Coincident Birthdays in a Group

A classic entertainment paradox in probability relates to the possibility of finding at least two people in a group who share the same birthday. In a group of 23 people, common sense is counterintuitive in that the probability of coincident birthdays is high.

	I	J	K
1	Repetition	Average Replay Utility 17.25%	Average of 20 Repetitions
2	1	4,036.98	4,031.47
3	2	4,277.08	
4	3	3,936.09	
5	4	4,246.59	Simulate
6	5	4,095.88	
7	6	3,870.08	
8	7	3,962.03	
9	8	4,103.11	
10	9	4,055.68	
11	10	4,161.16	
12	11	3,970.28	
13	12	4,104.76	
14	13	3,877.04	
15	14	4,025.04	
16	15	3,899.58	
17	16	4,004.13	
18	17	4,026.37	
19	18	4,077.69	
20	19	3,740.63	
21	20	4,159.21	

FIGURE 8.41   Vehicle insurance profits.

TABLE 8.7   Formulas for calculating profitability

Amount Insured Cell **B7**	=NORM.INV(RAND(); $B$2; $B$3)
Amount Charged **C7**	=(B7*$B$4)/100
Had an Accident? Y/N **D7**	= IF(RAND()<$B$1; "Y"; "N")
% to Compensate **E7**	=IF(D7="Y";LOOKUP(RAND();$C$3:$G$3;$D$1:$G$1);0)
Amount Paid by the Claim **F7**	=(B7*E7)/100
Utility **G7**	=C7-F7
Average Profit per Insured **H7**	=AVERAGE(G7:G10506)

Simulate to estimate the probability of finding in a gathering of 23 people, the probability that two or more people have the same birthday.

Solution

The following instructions program an Excel spreadsheet to estimate the probability that in a group of 23 people there are at least two people who have the same birthday.

From the Excel toolbar:

DEVELOPER / Macros / Create / Macro name / Birthdays

You can include this macro. In the same way, the button to execute the simulation is incorporated from toolbar:

DEVELOPER/ Insert / Form Controls / Button

And then bind the macro name to the control button.

```
Sub Birthdays()
 Dim A(23)
 Randomize
 Range("A1:F6").ClearContents
 Cells(1, "E") = "Run"
 Cells(1, "F") = "Probability"
 For Exec = 1 To 5
   Match = 0
   For Repetition = 1 To 100
     For i = 1 To 23
       A(i) = Int(365 * Rnd + 1)
     Next i
     For i = 1 To 6
       Cells(i, "A") = A(i)
       Cells(i, "B") = A(i + 6)
       Cells(i, "C") = A(i + 12)
     Next i
     For i = 19 To 23
       Cells(i - 18, "D") = A(i)
     Next i
     For X = 1 To 23
       U = A(X)
       For Y = X + 1 To 23
         If U = A(Y) Then Match = Match + 1
       Next Y
     Next X
   Next Repetition
   Cells(Exec + 1, "E") = Exec
   Cells(Exec + 1, "F") = Match / 10
 Next Exec
 Cells(7, "A") = "Probability(at least 2 coincident birthdays in 23 people):"
 Cells(7, "F") = "=AVERAGE(F2:F6)"
End Sub
```

	A	B	C	D	E	F
1	175	209	186	171	**Run**	**Probability**
2	124	307	214	239	1	0.7
3	360	9	363	294	2	0.77
4	19	253	196	154	3	0.65
5	284	174	196	293	4	0.78
6	365	163	306		5	0.63
7	**Probability of at least 2 coincident birthdays in a group of 23 people:**					**0.706**
8						
9		**Click to generate group of 23 people**				
10		**and their birthdays**				
11						

FIGURE 8.42   Matching birthdays.

At start the days of the year are coded from 1 to 365. Twenty-three random uniform values are generated between 1 and 365. Running the Excel macro as shown in Figure 8.36 generates the values in cells A1 through D6, but leaves cell D6 blank. From the macro several For/Next cycles at i and j take care of placing the uniform random values in Figure 8.42, cells A1:D6. These cells could also be filled directly to the sheet outside of the macro by means of the Excel function =RANDBETWEEN (1;365).

Subsequently, FOR/NEXT cycles in X and Y compare the 23 values with each other to determine the number of matching values. One FOR/NEXT cycle repeats 100 group simulations and another cycle runs ten repetitions. Figure 8.36 presents the simulation output. The probability of finding at least two people in a group of 23 who match on their birthday is estimated to be 0.706. The program performs a total of 1,000 simulations. The 23 birthdays of the last simulation appear in the output report in the range of cells A1:D6. There is a coincidence between cells C4 and C5; two people have a birthday on July 15, day 196.

Example 28: Random Sampling without Replacement
Sometimes it is required to obtain a series of random numbers without repetition, taken from a finite population list in order to use them in the random selection of individuals from the same population. If you have as input data the size of the population to be sampled and the number of values needed, simulate the generation of the indices that point to the positions of the data to be selected.

Solution
The following macro contains the Excel instructions to obtain a random list of indices used to select a sample of size M from a population P which can contain and represent up to 10,000 entities.

```
Sub Sampling()
   Dim A(10001)
   Cells(1, "A") = "Number of Entity to Sample"
   Range("A2:A10001").ClearContents
   P = Val(InputBox("How Big is your Population? (Max 10,000)", , 100))
   M = Val(InputBox("How many values do you want without replacement? ", , 10))
   Randomize
   A(2) = Int(P * Rnd + 1)
   For i = 1 To M
      Do
         U = Int(P * Rnd + 1)
      Repeat = False
      For j = 1 To i - 1
         If U = A(j) Then Repeat = True: Exit For
      Next j
      Loop While Repeat
      A(i) = U
   Next i
   For i = 1 To M
      Cells(i + 1, "A") = A(i)
   Next i
End Sub
```

Figure 8.43 gives the result of running the sampling macro to choose 10 values from a population of 100 pre-numbered and sorted values in a list.

Example 29: Solvent Bank Debtors
The historical records of a store's accounts receivable department show that its debtors fall into arrears at the exponential rate of 7% per year. Simulate 10,000

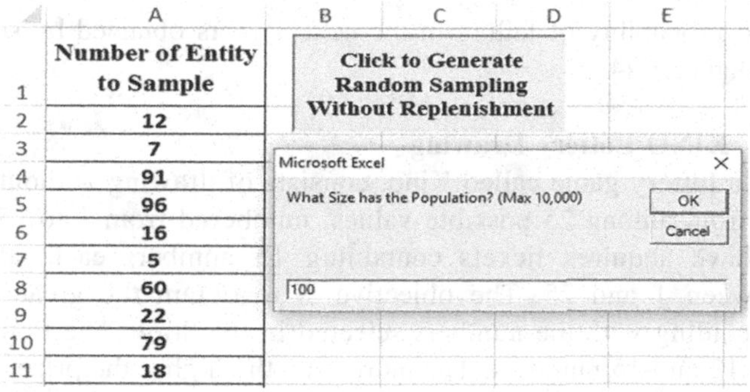

FIGURE 8.43 Sampling simulation.

	A	B	C	D	E
1	Debtor	Random	Duration time (years)		Prob(Solvent Debtor after 7 year)
2	1	0.5692	8.1		0.6113
3	2	0.2253	21.3		=COUNTIF(C2:C10001; ">7")/10000
4	3	0.0927	34.0		
5	4	0.3577	14.7		=-1/0,07*LN(B2)
6	5	0.1601	26.2		
7	6	0.6557	6.0		Prob(Debtor fails in 4 years)
8	7	0.4522	11.3		0.2495
9	8	0.4832	10.4		
10	9	0.2199	21.6		=COUNTIF(C2:C10001; "<=4")/10000

FIGURE 8.44   Bank debtors.

borrowers and determine: (A) Probability that a debtor remains solvent after 7 years. (B) Theoretical probability that a debtor of the bank will default in the next 4 years.

Solution
Figure 8.44, Cell C2 must be copied from cell C3 to cell C10001. For 10,000 debtors, cell E2 estimates the probability that a debtor will remain solvent after 7 years to be **0.6113**. The theoretical calculation is achieved by: $P(t > 7\ years) = e^{-0.07*7} = 0.6126$. The average of about 20 repetitions provides a better approximation.

The theoretical probability that a debtor will fail in the next 4 years is:

$$P(t \le 4\ \mathbf{years}) = 1 - e^{-0.07*4} = 0.2442$$

The estimated probability of failure in the next 4 years obtained by simulating was 0.2495. See Figure 8.44.

Example 30: KINO Lottery Drawing
The draw of a lottery game called Kino, consists of drawing without replacement 15 numbers from among 25 possible values, numbered from 1 to 25. Previously, the bettors have acquired tickets containing 15 numbers each, also randomly selected between 1 and 25. The objective is to obtain the greatest amount of numbers coinciding with the numbers selected in the draw. The Bettor wins if he hits between 12 and 15 numbers. The more hits, the higher the prize. Simulate this betting game.

	B	C	D	E	F	G	H	I	J	K	L
1	Gambler Card	Random for the Official Draw	Official Kino Card		Result of the Draw				Gambler Card		
2	1	0.8480	1		KINO.TACHIRA				SORTEO FECHA		(1315)
3	4	0.6469	5		1	5	8		1	4	5
4	5	0.7277	8		11	12	13		9	11	12
5	9	0.9735	11		14	15	16		13	14	15
6	11	0.3453	12		17	19	21		17	19	21
7	12	0.6354	13		22	24	25		22	23	24
8	13	0.2388	14								
9	14	0.1010	15				Hits		12		
10	15	0.2481	16			Parcial Winner 12 to 14					
11	17	0.5544	17								
12	19	0.6555	19			Click to Bet on Kino					
13	21	0.1612	21								
14	22	0.5764	22								
15	23	0.5841	24								
16	24	0.5995	25								

FIGURE 8.45   Simulation of the Kino drawing.

Solution

To simulate this game of chance it is necessary to generate independently and without replacement two series of 15 random numbers, where one series imitates a Kino card bet by a player and the other series the result of the official draw. It will be created a macro or subprogram named Play_Kino() which simulates the purchase of a card and the execution of the draw. This subprogram generates the contents of columns A, B, C, and D, as shown in Figure 8.45.

```
Sub Play_Kino()
   Randomize
   Dim C(25), S(25), CC(15), SS(15)
   'Generate 25 random numbers for the bettor's card C and 25 for the official
   draw S
   For I = 1 To 25
      C(I) = RND()
      S(I) = RND()
      Cells(I + 1, "A") = C(I)
```

```
    Cells(I + 1, "C") = S(I)
Next I
'Determine the place value for the first 15 numbers
For I = 2 To 16
    Cells(I, "B") = WorksheetFunction.Rank_Eq(Cells(I, "A"),
    Range("A2.A26"))
    Cells(I, "D") = WorksheetFunction.Rank_Eq(Cells(I, "C"),
    Range("C2.C26"))
Next I
'Assing the 15 numbers to the variables CC and SS
For I = 1 To 15
    CC(I) = Cells(I + 1, "B")
    SS(I) = Cells(I + 1, "D")
Next I
'Order the numbers from least to greatest on each card
For I = 1 To 15
    For J = I + 1 To 15
    If CC(I) > CC(J) Then
      Aux = CC(I)
      CC(I) = CC(J)
    CC(J) = Aux
    End If
  Next J
Next I
For I = 1 To 15
  For J = I + 1 To 15
  If SS(I) > SS(J) Then
      Aux = SS(I)
      SS(I) = SS(J)
      SS(J) = Aux
    End If
  Next J
Next I
'Write the ordered numbers in the cells of columns B and D.
    For I = 1 To 15
    Cells(I + 1, "B") = CC(I)
    Cells(I + 1, "D") = SS(I)
Next I
'Compare the numbers on the Bettor's card with the numbers in the official
Draw and count the hits.
  Hits = 0
  For I = 1 To 15
    For J = 1 To 15
      If CC(I) = SS(J) Then
```

```
            Hits = Hits + 1
         End If
       Next J
    Next I
    Range("I9").Value = Hits
 End Sub
```

Figure 8.45, row 1 range A1:K1, contains the column headings. The result of the simulation appears in range A2:D16.

Figure 8.46 shows expressions needed to mimic the results of the bettor's card and the draw.

Example 31: Bartlett's Test for Comparing Several Variances

Bartlett's test allows to determine if several variances from large samples can be assumed identical or not. This is a parametric test that assumes values from a normal distribution. Through simulation the efficiency of any statistical procedure to reject a null hypothesis can be evaluated.

Over a fully controlled experiment, Bartlett's test will be evaluated to determine its power to detect equality of variances present in four independent random samples each one of size 250 from normal populations of equal mean 300 and variance 30. Simulate 100,000 test to determine the proportion of rejections generated by applying Bartlett's test by using a confidence level of 95%.

Solution

Figure 8.47, cells in columns B2:B251, C2:C251, D2:D251, and E2:E251, contains the normal generating function Excel:

$$=NORM.INV(RAND();300;30)$$

The remaining cells are shown in Table 8.8.

	E	F	G	H	I	J	K
1	Result of the Draw				Gambler Card		
2							
3	=D2	=D3	=D4		=B2	=B3	=B4
4	=D5	=D6	=D7		=B5	=B6	=B7
5	=D8	=D9	=D10		=B8	=B9	=B10
6	=D11	=D12	=D13		=B11	=B12	=B13
7	=D14	=D15	=D16		=B14	=B15	=B16
8							
9			Hits	:	13		
10	=IF(I9=15; "Winner of 15 Hits "; IF(AND(I9>11; I9 <15); "Partial Winner 12 to 14 Hits"; "Less than 12 Hits"))						

FIGURE 8.46 Cells with their Excel formulas.

	A	B	C	D	E	F	G	H	I
1	Test	Sample 1	Sample 2	Sample 3	Sample 4		Barlett's Test for Variances Equality		Results
2	1	314.7	351.0	229.4	246.6		Variance 1	2140.35	Reject Ho
3	2	373.2	313.2	292.6	344.4		Variance 2	3104.65	
4	3	416.1	326.0	199.8	312.8		Variance 3	2211.66	Rejects
5	4	232.7	417.9	325.6	358.0		Variance 4	2772.21	4955
6	5	338.6	288.6	321.4	376.8		n	250	
7	6	352.1	301.0	267.8	289.6		N	1000	Percentaje of Rejects (%)
8	7	232.4	211.8	253.0	265.3		K	4	4.96
9	8	295.8	344.2	232.4	193.1		S_p^2	2557.22	
10	9	368.1	365.9	309.1	246.6		Log Var 1	3.33	Click to Simulate N
11	10	245.7	214.0	292.7	284.1		Log Var 2	3.49	Tests of Barlett
12	11	347.7	396.6	221.6	290.1		Log Var 3	3.34	
13	12	298.0	407.0	286.5	298.9		Log Var 4	3.44	
14	13	280.1	219.0	244.2	369.4		$Log(S_p^2)$	3.41	
15	14	371.0	336.2	310.4	400.6		Q	5.24	
16	15	325.7	377.1	285.5	280.5		h	1.00	
17	16	229.3	383.7	300.7	316.9		Barlett's Statistic	12.04	
18	17	359.9	333.1	342.3	285.7		Confidence Level	0.95	
19	18	208.1	311.4	296.2	363.7		Degrees of Freedom	3	
20	19	279.4	290.4	287.6	376.3		Chi-Squared	7.815	

FIGURE 8.47 Simulation of Bartlett's tests.

TABLE 8.8 Formulas Excel for Bartlett's

Cells H2:H5	=VAR.P(B2:B251) / VAR.P(C2:C251) / VAR.P (D2:D251) / VAR.P(E2:E251)	Calculate four sample variances
Cell H9	=((H6-1)/(H7-H8))*(H2+H3+H4+H5)	
Cells H10:H13	=LOG10(H2) / LOG10(H3) / LOG10(H4) LOG10(H5)	Get the logarithm of each variance
Cell H14	=LOG10(H9)	
Cell H15	=((H7-H8)*H14)-((H6-1)*(H10+H11+H12+H13))	Q
Cell H16	=1+((1/(3*(H8-1)))*((1/(H6-1))+(1/(H6-1))+(1/(H6-1)+(1/(H6-1)))))	
Cell H17	= (2.3026*H15)/H16	Determine Bartlett's statistic B
Cell H20	=CHISQ.INV(H18;H19)	
Cell I2	= IF(H17<H20;"NO Reject Ho";"Reject Ho")	

The content of the cells in column I5 and I8 are generated from the macro Sub Bartlett_Test (), shown below:

```
Sub Bartlett_Test()
    N = Val(InputBox("How Many Test to Run?", "Bartlett's Test / 4 Samples",
1000))
    R = 0
    Range("I5").ClearContents
    Range("I8").ClearContents
        For Test = 1 To N
            If Range("I2") = "Reject Ho" Then
                R = R + 1
            End If
            Cells(5, "I") = R
        Next Test
    Cells(4, "I") = "Rejects"
    Cells(7, "I") = "Percentage of Rejects (%)"
    Cells(8, "I") = (R / N) * 100
End Sub
```

In 100,000 tests, 4.96% of false positives or rejections were obtained. That is, 0.04% less than the theoretical expected value of 5% given that the confidence level declared in cell H18 is 95%.

Example 32: ANOVA Test for Four Samples

The ANOVA test is a generalization of the t-test albeit for comparing more than two groups of data. A simulation experiment will be performed to evaluate the ability of the ANOVA test to detect the equality of averages present in four sets of independent values, deliberately drawn from four randomly generated normal distributions with mean 150 and variance 5. The procedure to follow to apply the ANOVA test is described in Section 7.2.7 of chapter 7. It is desired to evaluate the goodness of fit of the ANOVA test to detect or not the equality of averages present in the four samples.

Solution

Four independent random samples of size 250 are generated from normal populations of equal mean 150 and variance 5. One hundred thousand ANOVA tests will be run to determine the proportion of rejections when not discovering the equality of averages present in the samples coming from populations certainly of equal average. The null hypothesis is:

Ho: The means of the normally distributed samples are statistically equal.

In Figure 8.48, the initial calculations to apply ANOVA are executed. Figures 8.49, 8.50, and 8.51 show the functions used in the calculations for Figures 8.48 and 8.52.

	B	C	D	E	F	G	H	I
	Average Sample 1	$\sum_{i=1}^{i=250}(Y_{i,1}-\overline{Y}_1)^2$	Average Sample 2	$\sum_{i=1}^{i=250}(Y_{i,2}-\overline{Y}_2)^2$	Average Sample 3	$\sum_{i=1}^{i=250}(Y_{i,3}-\overline{Y}_3)^2$	Average Sample 4	$\sum_{i=1}^{i=250}(Y_{i,4}-\overline{Y}_4)^2$
2	152.37	150,348.89	151.82	144,381.90	148.37	179,875.55	149.25	149,805.75
3								
4　n	$Y_{i,1}$	$(Y_{i,1}-\overline{Y}_1)^2$	$Y_{i,2}$	$(Y_{i,2}-\overline{Y}_2)^2$	$Y_{i,3}$	$(Y_{i,3}-\overline{Y}_3)^2$	$Y_{i,4}$	$(Y_{i,4}-\overline{Y}_4)^2$
5　1	164.64	150.45	200.43	2,363.43	152.68	18.64	144.99	18.16
6　2	154.13	3.10	153.87	4.22	131.80	274.54	178.13	834.05
7　3	179.91	758.40	176.27	597.71	153.91	30.70	132.92	266.88
8　4	151.45	0.85	136.94	221.49	132.04	266.61	118.85	924.22
9　5	164.67	151.29	165.86	197.10	105.55	1,833.32	160.26	121.28

FIGURE 8.48　ANOVA detection capability.

	A	B	C	D	E
1		Average Sample 1	$\sum_{i=1}^{i=250}(Y_{i,1}-\overline{Y}_1)^2$	Average Sample 2	$\sum_{i=1}^{i=250}(Y_{i,2}-\overline{Y}_2)^2$
2		=AVERAGE(B5:B254)	=SUM(C5:C254)	=AVERAGE(D5:D254)	=SUM(E5:E254)
3					
4	n	$Y_{i,1}$	$(Y_{i,1}-\overline{Y}_1)^2$	$Y_{i,2}$	$(Y_{i,2}-\overline{Y}_2)^2$
5	1	=NORM.INV(RAND();150;25)	=(B5-$B$2)^2	=NORM.INV(RAND();150;25)	=(D5-$D$2)^2
6	=A5+1	=NORM.INV(RAND();150;25)	=(B6-$B$2)^2	=NORM.INV(RAND();150;25)	=(D6-$D$2)^2
7	=A6+1	=NORM.INV(RAND();150;25)	=(B7-$B$2)^2	=NORM.INV(RAND();150;25)	=(D7-$D$2)^2
8	=A7+1	=NORM.INV(RAND();150;25)	=(B8-$B$2)^2	=NORM.INV(RAND();150;25)	=(D8-$D$2)^2
9	=A8+1	=NORM.INV(RAND();150;25)	=(B9-$B$2)^2	=NORM.INV(RAND();150;25)	=(D9-$D$2)^2

FIGURE 8.49　Formulas in columns A to E.

	F	G	H	I
1	Average Sample 3	$\sum_{i=1}^{i=250}(Y_{i,3}-\overline{Y}_3)^2$	Average Sample 4	$\sum_{i=1}^{i=250}(Y_{i,4}-\overline{Y}_4)^2$
2	=AVERAGE(F5:F254)	=SUM(G5:G254)	=AVERAGE(H5:H254)	=SUM(I5:I254)
3				
4	$Y_{i,3}$	$(Y_{i,3}-\overline{Y}_3)^2$	$Y_{i,4}$	$(Y_{i,4}-\overline{Y}_4)^2$
5	=NORM.INV(RAND();150;25)	=(F5-$F$2)^2	=NORM.INV(RAND();150;25)	=(H5-$H$2)^2
6	=NORM.INV(RAND();150;25)	=(F6-$F$2)^2	=NORM.INV(RAND();150;25)	=(H6-$H$2)^2
7	=NORM.INV(RAND();150;25)	=(F7-$F$2)^2	=NORM.INV(RAND();150;25)	=(H7-$H$2)^2
8	=NORM.INV(RAND();150;25)	=(F8-$F$2)^2	=NORM.INV(RAND();150;25)	=(H8-$H$2)^2
9	=NORM.INV(RAND();150;25)	=(F9-$F$2)^2	=NORM.INV(RAND();150;25)	=(H9-$H$2)^2

FIGURE 8.50　Formulas in columns F to I.

	K	L	M	N
1	Overall Mean		Items in each Group (n)	250
2	=(B2+D2+F2+H2)/4		Groups (k)	4
3			Sum of Squares Amongst Groups (SSA)	=N1*((B2-K2)^2+(D2-K2)^2+(F2-K2)^2+(H2-K2)^2)
4	Click to Simulate		Sum of Squares Within Groups (Error SSE)	=C2+E2+G2+I2
5			Total Sum of Squares (SST)	=N3+N4
6			Degrees of Freedom Amongst Groups (k-1)	3
7			Degrees of Freedom Within Groups (n-k)	246
8			Variance MSA	=N3/N6
9			Variance MSE	=N4/N7
10			Test Statistic F Computed	=N8/N9
11			Tabular F	=F.INV.RT(0.05;3;246)
12			Reject If Computed F is Greater than Tabular F	=IF(N10<N11;"Not Reject Ho";"Reject Ho")
13			Reject implies means of groups are different	
14			Reject Percentage (%)	

FIGURE 8.51 Formulas in columns K to N.

	J	K	L	M	N
1		Overall Mean		Items in each Group (n)	250
2		149.92		Groups (k)	4
3				Sum of Squares Amongst Groups (SSA)	7,036.25
4		Click to		Sum of Squares Within Groups (Error SSE)	630,187.04
5		Simulate		Total Sum of Squares (SST)	637,223.29
6				Degrees of Freedom Amongst Groups (k-1)	3
7				Degrees of Freedom Within Groups (n-k)	246
8				Variance MSA	2,345.42
9				Variance MSE	2,561.74
10				Test Statistic F Computed	0.92
11				Tabular F	2.64
12				Reject If Computed F is Greater than Tabular F	Not Reject Ho
13				Reject implies means of groups are different	
14				Reject Percentage (%)	0

FIGURE 8.52 ANOVA results.

Below are the commands necessary to repeat the execution of numerous ANOVA tests, included in the macro called Power_ANOVA:

```
Sub Power_ANOVA()
    Cells(14, "N").ClearContents
    N = Val(InputBox("How Many Evaluations To Run?", "ANOVA:
    4 Samples", 1000))
    R = 0
    For i = 1 To N
        If Range("N12") = "Reject Ho" Then
        R = R + 1
        End If
        Cells(14, "N") = R / N
    Next i
    Cells(14, "N") = (R / N) * 100
End Sub
```

From the results shown in Figure 8.52, we conclude that the ANOVA test capability for 100,000 runs for this example was 100%. The proportion of rejections was 0%. This is to be expected since all four samples were generated from normal distributions of equal mean and variance. This corresponds to evaluating the level of significance (α) of the test. It should be remembered that the significance level is the proportion of samples that are expected to be rejected even if they are true. In addition, ANOVA is a fairly robust test. When changing one of the four columns, for example, generating the first sample from an exponential distribution with an average of 200, a rejection of equality of means of 100% is detected.

It is left as an exercise to test the power of ANOVA if exponential values of average 200 are generated in column B maintaining the other three samples generated from two normal distributions of average 150 and variance 5. Use the expression =-200*LN(RAND()) in range B5:B254. Generate random samples from the distributions specified by the alternative hypothesis. Evaluate the test statistics from the simulated data and determine if the null hypothesis is accepted or rejected. The number of rejections is useful to calculate the power of the test.

Example 33: Aircraft Takeoff and Landing

The city's airport has one runway for aircraft landing and takeoff. The time between requests received by the control tower for takeoff or landing, is exponential random with an average of 20 minutes. The time taken for the maneuver is triangular, with values of 8 minutes minimum, 14 minutes maximum and most likely 10 minutes. Simulate the operation of runway service to airplanes to estimate the average waiting time. The airport has continuous operations.

	A	B	C	D	E	F	G	H	I J K
1	Airplane	Time Between Takeoff or Landing	Minute of Takeoff or Landing	Waiting Time (mins)	Minute of Start	Random Triangle	Duration of the Maneuver (mins)	Maneuver End Minute	a = 8
2	1	2.8	2.8	0	2.8	0.269	9.8	12.6	b = 10
3	2	0.3	3.1	9.5	12.6	0.786	13	25.6	c = 14
4	3	13.8	16.9	8.7	25.6	0.047	8.8	34.4	Z = 0.333
5	4	2.6	19.5	14.9	34.4	0.062	8.9	43.3	
6	5	8.7	28.2	15.1	43.3	0.172	9.4	52.7	Waiting Time
7	6	22.7	50.9	1.8	52.7	0.316	9.9	62.6	7.6
8	7	5.9	56.8	5.8	62.6	0.689	12.5	75.1	mins/airplane

FIGURE 8.53   Airport operation.

Solution

Figure 8.53 shows partial results obtained after simulating the service of 100,000 aircraft. To calculate the waiting time for each landing or takeoff, the Lindley equation was used:

$$W_q(n + 1) = Max\{0 \; ; \; W_q(n) + ST(n) - TBA(n + 1)\}$$

$W_q(n)$: Waiting time of the nth plane

$ST(n)$: Service Time of the nth plane

$TBA(n)$: Time Between Arrivals of the nth plane

displayed in the range D3:D100001, where the waiting time of the current entity (aircraft) $Wq(n+1)$ is equal to the greater value between 0 and the sum of the waiting time of the immediately preceding entity ($Wq(n)$, plus its service time $ST(n)$, minus the inter-arrival time of the current entity $TBA(n+1)$.

Figures 8.54, 8.55, and 8.56 show the formulas used to obtain the simulation results obtained in Figure 8.53.

In Figure 8.53, cell J7 shows the average waiting time for aircraft landing or taking off from the runway: 7.6 minutes/aircraft. About 20 runs from a macro allow a better estimate of the average aircraft waiting time; with its respective confidence interval. Table 8.9 contains the result of 20 independent runs of length 100,000 each.

	A	B	C	D	E	F
1	Airplane	Time Between Takeoff or Landing	Minute of Takeoff or Landing	Waiting Time (mins)	Minute of Start	Random Triangle
2	1	=ROUND(-20*LN(RAND()));1)	=B2	0	=C2	=ROUND(RAND(); 3)
3	=A2+1	=ROUND(-20*LN(RAND()));1)	=C2+B3	=ROUND(MAX(0; D2+G2-B3);1)	=MAX(H2;C3)	=ROUND(RAND(); 3)
4	=A3+1	=ROUND(-20*LN(RAND()));1)	=C3+B4	=ROUND(MAX(0; D3+G3-B4);1)	=MAX(H3;C4)	=ROUND(RAND(); 3)

FIGURE 8.54   Formulas in columns A to F.

	G
1	Duration of the Maneuver (mins)
2	=ROUND(IF(F2<$K$4; $K$1+($K$3-$K$1)*SQRT($K$4*F2); $K$1+($K$3-$K$1)*(1-SQRT(1-$K$4)*(1-F2)));1)
3	=ROUND(IF(F3<$K$4; $K$1+($K$3-$K$1)*SQRT($K$4*F3); $K$1+($K$3-$K$1)*(1-SQRT(1-$K$4)*(1-F3)));1)
4	=ROUND(IF(F4<$K$4; $K$1+($K$3-$K$1)*SQRT($K$4*F4); $K$1+($K$3-$K$1)*(1-SQRT(1-$K$4)*(1-F4)));1)

FIGURE 8.55   Formulas in column G.

	H	I	J	K
1	Maneuver End Minute		a =	8
2	=E2+G2		b =	10
3	=E3+G3		c =	14
4	=E4+G4		Z =	=(K2-K1)/(K3-K1)
5	=E5+G5			
6	=E6+G6		Waiting Time	
7	=E7+G7		=AVERAGE(D2:D100001)	
8	=E8+G8		mins/airplane	

FIGURE 8.56   Formulas in columns H to K.

TABLE 8.9   Result of 20 runs

7.6	7.47	7.53	7.84
.61	7.66	7.65	7.73
7.69	7.64	7.51	7.68
7.63	7.77	7.59	7.66
7.58	7.54	7.62	7.58

The average time and its deviation for the twenty values is 7.629 and 0.0883, respectively. The Excel statistical function:

=CONFIDENCE.NORM(0.01;0.0883;20) = 0.05086

Allows us to establish a 99% confidence interval for the average aircraft waiting time:

[7.58 ≤ *Waiting Time* ≤ 7.67] minutes/airplane

Example 34: TV Customers Served by Two Employees
The customer service of a cable TV company is attended by two employees A and B. A is preferred by the users since he manages to serve a Poisson average of 15 users per hour, while B only 12. People arrive according to a Poisson distribution with an

average of 20 per hour. Simulate serving 10,000 users and determine the most important performance statistics of the system. The company compensates each customer who has to wait for service with a $5 bonus. Estimate the expected amount to be disbursed for bonuses during a daily 12-hour service day.

Solution

To estimate system and waiting times, it is required to individualize the arrival and departure times of each user. Therefore, it is necessary to transform the Poisson averages into exponential averages that reflect the time between arrival and service, preferably in minutes:

Poisson Arrival rate: $\lambda = 20$ users/(60 min.) <==> Exponential Time Between Arrivals: $1/\lambda = 3$ min./user

Poisson Service Time: $\mu_A = 15$ users/(60 min.) <==> Exponential service time: $1/\mu_A = 4$ min./user

Poisson Service Time: $\mu_B = 12$ users/(60 min.) <==> Exponential service time: $1/\mu_B = 5$ min./user

The input values: inter-arrival and service time, will be arranged in the range B2:B5, as shown in Figure 8.57. Ten thousand users placed in A7:A10006 will be simulated with their respective inter-arrival times.

In order to simplify the model by avoiding inter-arrival times of less than one minute, the exponential random generator is included as an argument of the =MAX () function. See Figure 8.58, cell B7. So that the minimum possible time between the arrival of one customer and the next corresponds to at least one minute.

In column C, range C8:C10006, to find the exact time of arrival of a client, the inter-arrival time (TBA) is added to the time of arrival of the preceding user. Columns D to F present a logical procedure to select employee A or B according to their availability. See Figure 8.58.

	Input Data			Arrival/Service Rates		Performance measures in only one simulation run					
	Time Between Customer Arrivals (TBA)	3	mins/client	λ (20 clients/hour)	Mean Time in System W =	8.8	mins/client	Average Clients in System L =	2.9	customers	Probability that a client has to wait
	Mean service time (A)	4	mins/client	μ_A (15 clients/hour)	Wait time Wq =	4.7	mins/client	Queue Length Lq =	1.6	customers	0.5552
	Mean service time (B)	5	mins/client	μ_B (12 clients/hour)							

Customer n	TBA(n) (mins)	Arrival at Time T(n) (min)	Server A available at time	Server B available at time	Chosen Server	Attention Time (mins)	Service Begins at (min)	Server A End at Time (min)	Server B End at Time (min)	Waiting Time Wq(n) (mins)	System Time W(n) (mins)
1	1	1	0	0	A	11	1	12		0	11
2	3	4	12	0	B	1	4	12	5	0	1
3	1	5	12	5	B	4	5	12	9	0	4
4	3	8	12	9	B	8	9	12	17	1	9
5	1	9	12	17	A	2	12	14	17	3	5
6	2	11	14	17	A	8	14	22	17	3	11
7	1	12	22	17	B	10	17	22	27	5	15
8	2	14	22	27	A	1	22	23	27	8	9
9	1	15	23	27	A	7	23	30	27	8	15

FIGURE 8.57   Service desk simulation.

	A	B	C	D	E	F
1		Input Data			Arrival/Service Rates	
2	Time Between Customer Arrivals (TBA)	3	mins/client	λ (20 clients/hour)		Mean Time in System W =
3	Mean service time (A)	4	mins/client	μ_A	(15 clients/hour)	Wait time Wq =
4	Mean service time (B)	5	mins/client	μ_B	(12 clients/hour)	
5						
6	Customer n	TBA(n) (mins)	Arrival at Time T(n) (min)	Server A available at time (min)	Server B available at time (min)	Chosen Server
7	1	=MAX(1;INT(-$B$2*LN(RAND())))	=B7	0	0	A
8	=A7+1	=MAX(1;INT(-$B$2*LN(RAND())))	=C7+B8	=I7	0	=IF(D8<=E8;"A";"B")
9	=A8+1	=MAX(1;INT(-$B$2*LN(RAND())))	=C8+B9	=IF(I8<=C9;C9;I8)	=IF(J8<=C9;C9;J8)	=IF(D9<=E9;"A";"B")

FIGURE 8.58 Formulas used in the service desk.

Columns G and H simulate the service time and the service start time (Figure 8.59). Also for simplicity, the MAX() function guarantees a service time greater than or equal to one minute. The range G2:G3 allows estimating the average time in the system (W) and the waiting time (Wq) in minutes for all customers.

Figure 8.60 provides the relationships for estimating L: the average number of customers in the office at any instant and Lq: the average length of customers waiting. This calculation is based on Little's equations, where:

$$L = W * \lambda \text{ and } Lq = Wq * \lambda$$

	G	H
1		Performance measures in only one simulation ru
2	=AVERAGE(L7:L10006)	mins/client
3	=AVERAGE(K7:K10006)	mins/client
4		
5		
6	Attention Time (mins)	Service Begins at (min)
7	=MAX(1;INT(-B3*LN(RAND())))	=C7
8	=IF(F8="A"; MAX(1;INT(-$B$3*LN(RAND()))); MAX(1;INT(-$B$4*LN(RAND()))))	=C8
9	=IF(F9="A"; MAX(1;INT(-$B$3*LN(RAND()))); MAX(1;INT(-$B$4*LN(RAND()))))	=IF(F9="A";MAX(C9;D9);MAX(C9;E9))
10	=IF(F10="A"; MAX(1;INT(-$B$3*LN(RAND()))); MAX(1;INT(-$B$4*LN(RAND()))))	=IF(F10="A";MAX(C10;D10);MAX(C10;E10))

FIGURE 8.59 Functions for service time.

	I	J	K	L
1	nulation run			
2	Average Clients in System L =	=(1/B2)*G2	customers	Probability that a client has to wait
3	Queue Length Lq =	=(1/B2)*G3	customers	=COUNTIF(K7:K10006;">0")/10000
4				
5				
6	Server A End at Time (min)	Server B End at Time (min)	Waiting Time Wa(n) (mins)	System Time W(n) (mins)
7	=H7+G7		0	=I7-C7
8	=IF(F8="A";D8+G8;I7)	=IF(F8="B";G8+H8;J7)	=H8-C8	=IF(F8="A";I8-C8;J8-C8)
9	=IF(F9="A";G9+H9;IF(C9<=I8;I8;C9))	=IF(F9="B";G9+H9;J8)	=H9-C9	=IF(F9="A";I9-C9;J9-C9)
10	=IF(F10="A";G10+H10;IF(C10<=I9;I9;C10))	=IF(F10="B";G10+H10;J9)	=H10-C10	=IF(F10="A";I10-C10;J10-C10)

FIGURE 8.60 Formulas for other performance indicators.

To estimate the total amount to be compensated during a 12-hour workday, it is indispensable to obtain the probability that a customer will have to wait. This value multiplied by the number of people expected to arrive during 12 hours, allows us to estimate the number of people waiting during the day. Since Figure 8.57, range K7:K10006 contains the waiting times, the probability of not waiting arises from counting the number of cells containing zero waiting values and dividing that total by 10,000 customers.

The average number of customers arriving per hour is known at the beginning with value λ = 20 customers/hour. In 12 hours, 240 people are expected to arrive. The probability of a customer waiting was estimated at 0.5552. That is, of the 240 customers that on average arrive during a 12-hour day, it is estimated that on average 133 customers will wait (240 * 0.5552), at an individual cost of $5. Consequently, the total expected amount of waiting bonus is approximately $665.

Example 35: Seligman's Happiness Equation
Martin Seligman, pioneer of the so-called positive psychology, in 1996 presented his equation for estimating the level of individual happiness **H** expressed by accumulating the contribution of three factors:

$$H = G + CL + PW$$

Where:

G:	A genetic value whose contribution is less than 15%.
CL:	Circumstances of life with a contribution between 8% and 15%.
PW:	Factors dependent on personal will. At least 50% of the level of happiness.

Solution
Figure 8.61 shows the result of the simulation of the happiness equattion.

	A	B	C	D	E	F	G
1	Person	Genetic Range G:	Circumstances of Life CL	Personal Willness PW	Happiness Level		
2	1	3	15	58	76		→=B2+C2+D2
3	2	14	13	60	87		
4	3	3	10	52	65		
5	4	15	8	56	79		
6	5	15	8	66	89		
7		↓ =RANDBETWEEN(1;15)	↓ =RANDBETWEEN(8;15)	↓ =RANDBETWEEN(50; 100-(B2+C2))			

FIGURE 8.61 Level of happiness according to Seligman.

Example 36: Rural Aqueduct

A reservoir can store up to 50,000 liters of water. Every 6 hours it supplies a random amount of the liquid to an urban piped network, according to a uniform distribution with parameters between 3,000 and 6,000 liters. However, it could deliver less or none depending on availability. The reservoir only collects rainwater. Rainfall in the area occurs according to a Poisson distribution with an average of 12 rains per week. The amount of water that could enter the reservoir during a downpour is random normal with an average of 9,000 and a deviation of 2,500 liters. For simplicity, we approximate the times of occurrence of rainfall to whole hour values. The reservoir starts full. Discharge and rainfall events can coincide at the same time or hour. (A) Determine the simulation length to estimate the average amount of the reservoir contents at any instant, at the 99% $(1 - \alpha)$ confidence level and 500 liter (ε) error. (B) Estimate the expected contents of the reservoir at any time of the day. (C) What is the probability that the reservoir will be empty?

Solution

The input data and informative texts are arranged in the range A2:G6. See Figure 8.62. The time between rains is calculated in cell F6 by the relation: =(7*24)/(12), which yields 14 hours/rainfall.

Initially the simulation area is defined in the range A9:L5008. A pilot sample of 5,000 observations will be used to estimate the simulation length. The variable of interest is the level of the reservoir at any instant, column L. A histogram was constructed with the 5,000 values of the sample and its statistics was observed: median, average and mode. Since it does not present a normal distribution behavior, to estimate the size of the simulation the Chebyshev relation will be applied according to:

$$n \geq \frac{s^2}{\alpha * \varepsilon^2} = \frac{205,080,483.4}{0.01 * 500^2} \cong 82,032$$

To determine the sample size, cell I3 contains the ratio =J6/(0.01*500*500). If confidence level is $(1 - \alpha) = 0.99$, *then* $\alpha = 0.01$. Given the high variability of the

	A	B	C	D	E	F	G	H	I	J	K	L
1	Input Data								Experiment Responses			
2	Deposit Capacity:	50,000 liters				Average	Deviation		A) Length	B) Content	Times the Tank is Empty	C) Probability.
3		Min	Max		Input Rain:	9,000	2,500 liters		82,032	16,549	14,804	0.164
4	Volume to Download every 6	3,000	6,000 liters									
5												
6	Average Rainfall (Poisson):	12	rains/week	<====>	Time Between Rains (Exp):	14	hours/rain (Exp.)		Deposit Level Variance	205,080,483.4		
7												
8	Time (hour)	Deposit Level at the Beginning (liters)	Random Next Rain R	Time Between Rains (hours)	Time the Next Rain R Will Fall	Time the Next Download D Will Occur	Random to Download Estimate	Amount to Download (liters)	Amount to Enter the Deposit	Next Event R/D/ B (Both)	Deposit Amount Discharged	Deposit Level at End of Hour
9	0	50,000	0.5422	9	9	6	--	--	--	D	--	50,000
10	6	50,000	--	--	9	12	0.419	4,256	-	R	4,256	45,744
11	9	45,744	0.7399	5	14	12	--	-	6,873	D	-	50,000
12	12	50,000	--	--	14	18	0.783	5,347	-	R	5,347	44,653
13	14	44,653	0.9646	1	15	18	--	-	5,437	R	-	50,000
14	15	50,000	0.8234	3	18	18	--	-	10,276	B	-	50,000
15	18	50,000	0.2091	22	40	24	0.666	4,997	7,904	D	4,997	50,000

FIGURE 8.62   Inputs and results.

reservoir level, the experiment will be carried out for 90,000 events, almost 10% above the calculated n value.

Figures 8.63, 8.64, 8.65, and 8.66 show the formulas necessary to simulate and achieve the results presented in Figure 8.62.

The variability of reservoir content level s^2 is high. See Figure 8.62, Cell J6. It increases the experiment length. After 90,000 events were simulated, range I1:L3,

	A	B	C	D
1				Input Data
2	Deposit Capacity:	50000	liters	
3		Min	Max	
4	Volume to Download every 6 hours:	3000	6000	liters
5				
6	Average Rainfall (Poisson):	12	rains/week	<====>
7				
8	Time (hour)	Deposit Level at Beginning (liters)	Random For Next Rain	Time Between Rains (hours)
9	0	=$B$2	=RAND()	=INT(-$F$6*LN(C9))+1
10	=MIN(E9;F9)	=L9	=IF(OR(J9="R"; J9= "B"); RAND();"--")	=IF(OR(J9="R";J9="B"); INT(-$F$6*LN(C10))+1; "--")
11	=MIN(E10;F10)	=L10	=IF(OR(J10="R"; J10= "B"); RAND();"--")	=IF(OR(J10="R";J10="B"); INT(-$F$6*LN(C11))+1; "--")

FIGURE 8.63 Formulas used for rain fall simulation.

	E	F	G
1			
2		Average	Deviation
3	Input Rain:	9000	2500
4			
5			
6	Time Between Rains (Exp):	=7*24/(12)	hours/rain (Exp.)
7			
8	Time the Next Rain R Will Fall	Time the Next Download D Will Occur	Random to Estimate Download
9	=+D9	6	--
10	=IF(OR(J9="R";J9="B"); E9+D10;E9)	=IF(OR(J9="D";J9="B"); F9+6;F9)	=IF(OR(J9="D";J9="B"); RAND(); "--")
11	=IF(OR(J10="R";J10="B"); E10+D11;E10)	=IF(OR(J10="D";J10="B"); F10+6;F10)	=IF(OR(J10="D";J10="B"); RAND(); "--")
12	=IF(OR(J11="R";J11="B"); E11+D12;E11)	=IF(OR(J11="D";J11="B"); F11+6;F11)	=IF(OR(J11="D";J11="B"); RAND(); "--")

FIGURE 8.64 Formulas used in columns E, F, and G.

	H	I
1		
2		A) Length
3	liters	=J6/(0.01*500*500)
4		
5		
6		Deposit Level Variance
7		
8	Amount to Download (liters)	Amount to Enter the Deposit
9	--	--
10	=IF(OR(J9="D";J9="B"); INT($B$4+ ($C$4-$B$4)*G10); 0)	=IF(OR(J9="R";J9="B"); INT(NORM.INV(RAND();$F$3;$G$3)); 0)
11	=IF(OR(J10="D";J10="B"); INT($B$4+ ($C$4-$B$4)*G11); 0)	=IF(OR(J10="R";J10="B"); INT(NORM.INV(RAND();$F$3;$G$3)); 0)
12	=IF(OR(J11="D";J11="B"); INT($B$4+ ($C$4-$B$4)*G12); 0)	=IF(OR(J11="R";J11="B"); INT(NORM.INV(RAND();$F$3;$G$3)); 0)

FIGURE 8.65 Formulas used in columns H and I.

	J	K	L
1	Experiment Responses		
2	B) Content	Times the Tank is Empty	C) Probability.
3	=AVERAGE(L9:L90008)	=COUNTIF(L9:L90008;0)	=K3/90000
4			
5			
6	=VAR.S(L9:L5008)		
7			
8	Next Event R/D/ B (Both)	Deposit Amount Discharged	Deposit Level at End of Hour
9	=IF(E9>F9; "D";IF(E9=F9; "B"; "R"))	--	=B9
10	=IF(E10>F10; "D";IF(E10=F10; "B"; "R"))	=IF(AND(H10>0; B10>H10); H10; IF(H10=0;0;B10))	=MIN($B$2;B10-K10+I10)
11	=IF(E11>F11; "D";IF(E11=F11; "B"; "R"))	=IF(AND(H11>0; B11>H11); H11; IF(H11=0;0;B11))	=MIN($B$2;B11-K11+I11)
12	=IF(E12>F12; "D";IF(E12=F12; "B"; "R"))	=IF(AND(H12>0; B12>H12); H12; IF(H12=0;0;B12))	=MIN($B$2;B12-K12+I12)

FIGURE 8.66   Excel formulas used in columns J, K, and L.

Figure 8.62 shows answers. In Figure 8.62, cell J3, the average content of the reservoir at any time of the day is estimated as 16,549 liters. Moreover, the probability to find the reservoir empty is 0.164. To get better estimated values, it could be useful to average other 19 more runs. However, since the simulation length was calculated to an error of 500 liters, then with 99% confidence, the average volume can be expected to be between 16,049 and 17,049 liters. The estimator plus or minus 500.

Example 37: River and Dam[*]
The national government plans to build a dam in order to: (A) control flooding in the area, (B) provide irrigation for the farming community, and (C) provide recreational activities. The maximum capacity of the dam would be 4 million cubic meters of water. One million cubic meters corresponds to one unit. The weekly flow in millions of cubic meters of water from the river that will supply the dam is random but historically fits the following possibilities:

Weekly Flow (units)	2	3	4	5	6	7
Probability	0.10	0.15	0.35	0.20	0.10	0.10

The government must fulfill an irrigation contract in the area in the amount of two units of water per week; however, if it is not enough, it can deliver only one unit. Additionally, to maintain water quality it is mandatory to release one unit of water per week. Consequently, the normal weekly water release target is three units. However, when the dam level plus inflow is less than the three units indicated above, the shortage must be resolved at the expense of water dedicated to irrigation. If the dam is full, any additional inflow shall be released immediately. Ecological policy prohibits reducing the dam level to less than one unit. The weekly revenue for servicing the irrigation agreement is $1.5 million/unit. If it is not possible to meet the irrigation supply, the government must compensate at the cost of $2 million per unit neglected. Revenues for such purposes depend on the level of the dam. In thousands, revenues are distributed as follows:

Level	1	2	3	4
Revenues ($)	0	4,000	6,000	2,000

Finally, if it is necessary to over-release 2 or more units of water, a flood loss of $5 million per unit is incurred. Simulate 50,000 weeks and determine: (A) Expected weekly profit. (B) What is the probability of flooding? (C) How many floods are expected to occur during a year? *This exercise, although solved by simulation, is inspired by the Markov chain model proposed in the book "Introduction to Operation Research Techniques".

Solution

In this model we will assume the inflow of all events of inflow or outflow of liquid units concurrently taken at the beginning of each week. That is, without considering the actual capacity constraint, at the beginning of each week the stock and river flow are summed and then the corresponding outflow deductions are deducted to determine the final balance for the week.

In Figure 8.67, the cells of the range A1:L8 contain the legends, input data and shows experiment responses in the range J4:L6.

Cell B9 contains the expression =B1 corresponding to the dam capacity. Then cell B10 must contain the expression =G9 to bring to the beginning the final level of the previous week.

The weekly water flow of the river is simulated by the expression:

=LOOKUP(RAND();$A$6:G$6;B$4:G$4)

Where the random value is searched in the cumulative probability range A6:G6 and its flow value in the range B4:G4.

	A	B	C	D	E	F	G	H	I	J	K	L
1	Dam Capacity:	4		Irrigation Income (thousands)	Fine for Failure to Comply	Flood Penalty		Dam Level (units):	1	2	3	4
2	Irrigation Contract:	2		1,500	2,000	5,000		Recreation Income:	0	4,000	6,000	2,000
3	Output by Quality:	1										
4	Weekly flow:	1	2	3	4	5	6	7		Experiment Responses		
5	Probability:	0.15	0.30	0.20	0.15	0.10	0.05	0.05		A) E(Utility)	B) P(Flood)	C) E(Floods)
6		0	0.15	0.45	0.65	0.8	0.9	0.95	1	3,594.77	0.088	4.6
7												
8	Week	Start Level	Weekly Flow	Total Weekly to Assign	Irrigation	Units to Release due to Excess	Final Level	Fine for Failure to Comply Irrigation	Weekly Irrigation Rent	Recreation Income	Weekly Penalty for Flooding	Weekly Profit (thousands $)
9	1	4	4	8	2	1	4	-	3,000.00	2,000.00	-	5,000.00
10	2	4	1	5	2	0	2	-	3,000.00	4,000.00	-	7,000.00
11	3	2	2	4	2	0	1	-	3,000.00	-	-	3,000.00
12	4	1	2	3	1	0	1	2,000.00	1,500.00	-	-	-500.00
13	5	1	4	5	2	0	2	-	3,000.00	4,000.00	-	7,000.00
14	6	2	7	9	2	2	4	-	3,000.00	2,000.00	10,000.00	-5,000.00

FIGURE 8.67 Dam simulation results.

In cell D9, to obtain the total availability of fluid to be handled, the existing level and the random amount flowing in the week are accumulated by =B9+C9.

In cell E9 it is decided whether the stock exceeds two units to meet the irrigation. It should be remembered that the reservoir cannot be empty and must also empty one unit to maintain its quality. The nested IF expression:

=IF(D9>2; IF(D9=3;1;2);0)

Decides whether to deliver for irrigation none, one, or two units.

Cell F9 contains =IF(D9>7;D9-7;0) to decide the number of units to release. If the stock is greater than 7 units, the quantity exceeding 7 must be released.

The stock at the end of each week is given by the expression:

=IF(D9>7;$B$1;D9-E9-1)

Where it is checked if the total quantity to be allocated exceeds 7 units, which would leave the reservoir full to its maximum capacity. Otherwise, the balance corresponds to deducting the amount served for irrigation and the unit for water quality control.

The expressions to determine the penalty for non-compliance and irrigation income:

=(2-E9)*$E$2 and E9*$D$2.

Weekly recreation revenue is achieved by the expression:

=LOOKUP(G9;I$1:L$1;$I$2:$L$2).

The flooding penalty is estimated from: =IF(F9>=2;$F$2*F9;0) in cell K9.

The weekly economic benefit in cell L9 by relating income minus outflows is defined by the following expression:

=I9 + J9 − H9 − K9

Finally, the range of row 9, C9:L9, must be copied over the range of row 10, C10:L10; and then copy all of row 10, range A10:L10 into the range A11.L50008.

The answers will appear in cells J5, K5, and L5.

- A. The expected weekly utility is obtained from the average of column L:
  =AVERAGE(L9:L50008)
  In average, utility is expected to be $3,594,770 per week.
- B. The probability of flooding, cell K6, is obtained by counting the number of times a flooding penalty is paid and dividing by 50,000 weeks:
  =COUNTIF(K9:K50008;"<>0")/50000
  The probability of having a flood in any week is estimated to be 0.088.

C. The expected number of annual floods is obtained by multiplying 52 weeks of a year by the probability of a weekly flood occurring: =52*K6. It is expected to have about 4.6 floods/year.

Running about 20 repetitions would improve the accuracy for each of the probable values (A), (B), and (C). All input parameters can be modified. For example, capacity can be taken as a decision variable to evaluate various scenarios and select the most profitable one.

Figure 8.68 and 8.69 show the formulas used to carry out the simulation.

Example 38: Spare Parts to Attend Failures of a Pneumatic System

A pneumatic system contains an electronic device whose failure can be assumed to be a Poisson average of 0.0001 failure/hour. Any failure is corrected immediately using a spare unit whose lifetime is independent without the system suffering a downtime. It is required to determine the number of spare units that must be had to ensure that 95% of the time the system is functional for at least 30,000 hours. Simulate 50,000 pneumatic systems.

Solution

For a given number of spare drives, the time to system failure needs to be evaluated. It will start with two spare units and the probability of failure will be calculated for 50,000 systems. Then the number of spare units will be increased until a fidelity of 95% or more is obtained. The proposed model corresponds to an Erlang(S, 0.0001) distribution (Figure 8.70).

	A	B	C	D	E	F	G
8	Week	Start Level	Weekly Flow	Total Weekly to Assign	Irrigation	Units to Release due to Excess	Final Level
9	1	=B1	=LOOKUP(RAND();$A$6:G$6;B$4:G$4)	=B9+C9	=IF(D9>2; IF(D9=3;1;2);0)	=IF(D9>7;D9-7;0)	=IF(D9>7;$B$1;D9-E9-1)
10	=A9+1	=G9	=LOOKUP(RAND();$A$6:G$6;B$4:G$4)	=B10+C10	=IF(D10>2; IF(D10=3;1;2);0)	=IF(D10>7;D10-7;0)	=IF(D10>7;$B$1;D10-E10-1)
11	=A10+1	=G10	=LOOKUP(RAND();$A$6:G$6;B$4:G$4)	=B11+C11	=IF(D11>2; IF(D11=3;1;2);0)	=IF(D11>7;D11-7;0)	=IF(D11>7;$B$1;D11-E11-1)

FIGURE 8.68   Formulas for dam simulation A1:G50008.

	H	I	J	K	L
5	0.05		A) E(Utility)	B) P(Flood)	C) E(Floods)
6	=G6+H5		=AVERAGE(L9:L50008)	=COUNTIF(K9:K50008;"<>0")/50000	=K6*52
7					
8	Fine for Failure to Comply Irrigation	Weekly Irrigation Rent	Recreation Income	Weekly Penalty for Flooding	Weekly Profit (thousa
9	=(2-E9)*$E$2	=E9*$D$2	=LOOKUP(G9;I$1:L$1;$I$2:$L$2)	=IF(F9>=2;$F$2*F9;0)	=I9+J9-H9-K9
10	=(2-E10)*$E$2	=E10*$D$2	=LOOKUP(G10;I$1:L$1;$I$2:$L$2)	=IF(F10>=2;$F$2*F10;0)	=I10+J10-H10-K10
11	=(2-E11)*$E$2	=E11*$D$2	=LOOKUP(G11;I$1:L$1;$I$2:$L$2)	=IF(F11>=2;$F$2*F11;0)	=I11+J11-H11-K11

FIGURE 8.69   Formulas for dam experiment H5:L50008.

	A	B	C	D	E	F	G	H	I	J	K
1	Input Data									Results	
2	K	E(x) = 1/λ								Components Fail After 30,000 hours	Estimated Probability
3	7	10000								48364	0.9673
4											
5	System	Random R1	Random R2	Random R3	Random R4	Random R5	Random R6	Random R7		R1*R2*R3*R4*R5*R6*R7	Time to Failure (hours)
6	1	0.376	0.579	0.003	0.624	0.190	0.434	0.817		0.00002	106,085.8
7	2	0.267	0.634	0.046	0.179	0.753	0.053	0.770		0.00004	100,401.6
8	3	0.445	0.693	0.667	0.875	0.333	0.485	0.079		0.00229	60,810.0
9	4	0.813	0.041	0.243	0.855	0.038	0.058	0.428		0.00001	119,313.5
10	5	0.083	0.373	0.798	0.383	0.746	0.412	0.061		0.00018	86,427.6
11	6	0.996	0.720	0.186	0.208	0.902	0.454	0.219		0.00249	59,935.8
12	7	0.951	0.327	0.623	0.012	0.523	0.733	0.850		0.00074	72,154.4
13	8	0.920	0.135	0.714	0.670	0.873	0.215	0.661		0.00735	49,127.8
14	9	0.350	0.788	0.375	0.283	0.901	0.888	0.871		0.02033	38,955.9

FIGURE 8.70   Simulation and results.

Summary of scenarios:

If K Spare Units are:	2	3	4	5	6	7
Failure Probability:	0.199	0.420	0.645	0.811	0.917	0.967

Figure 8.71 shows formulas used in Figure 8.70.
More than 95% confidence is achieved by having in stock 7 spare units (96.7%).

Example 39: Triple Two Dice
A gambler rolls two dice. On each roll he bets $100. If the player obtains the same value on both dice, he receives three times the amount wagered. He starts with $1,000. Since the player bets to his undoing, determine the number of bets he expects to place until he runs out of funds (A) analytically and (B) by simulating.

Solution

A. Analytically: It is required to determine the income expectation in any bet. The probability of getting both heads the same and make a $200 profit is to have

	A	B	C	D	E	F	G	H	I	J	K
1	Input Data									Results	
2	K	E(x) = 1/λ								Components Fail After 30,000 hours	Estimated Probability
3	7	=1/0.0001								=COUNTIF(K6:K50005;">30000")	=J3/50000
4											
5	Component	Random R1	Random R2	Random R3	Random R4	Random R5	Random R6	Random R7		R1*R2*R3*R4*R5*R6*R7	Time to Failure (hours)
6	1	=RAND()	=RAND()	=RAND()	=RAND()	=RAND()	=RAND()	=RAND()		=B6*C6*D6*E6*F6*G6*H6	=-$B$3*LN(J6)
7	=A6+1	=RAND()	=RAND()	=RAND()	=RAND()	=RAND()	=RAND()	=RAND()		=B7*C7*D7*E7*F7*G7*H7	=(-$B$3)*LN(J7)

FIGURE 8.71   Formulas for failures.

6 favorable outcomes out of 36 possible outcomes. That is, 1/6. On the contrary, the probability of losing a $100 bet is 30 outcomes against 36 possible outcomes. That is to say, 5/6. Consequently, the expected income is calculated as:

$$E\,(Profit) = \left(200 * \frac{1}{6}\right) - \left(100 * \frac{5}{6}\right) = \$ -50$$

Which means that a bettor expects to lose $50 for each bet placed. Since your available betting capital is $1,000, then you are expected to place 1,000/50 = 20 bets until you are broke. This result will be estimated by simulation.

B. By simulation

Figure 8.72 shows the result of performing 10,000 repetitions. It is observed that a gambler expects to place about 20 bets before going broke.

The Win_3 macro program allows you to carry out the simulation for any amount of desired number of repetitions. A repetition corresponds to a series of bets until the bettor loses all his initial capital. The procedure for using Excel macro programs is detailed in Sections 2.14 to 2.18 of Chapter 2.

	A	B	C	D	E	F
1	Initial capital	Amount placed in each bet	Prize factor	How many repetitions for the experiment?		Amount of bets until all the capital is lost
2	1,000.00	100.00	3	10000		20.10
3						
4	Bet	Die 1	Die 2	Credit available		
5	1	3	5	900.00		Click
6	2	2	3	800.00		to Roll 2 Dice
7	3	5	6	700.00		
8	4	1	1	900.00		
9	5	1	5	800.00		
10	6	2	1	700.00		
11	7	3	3	900.00		
12	8	5	6	800.00		
13	9	5	4	700.00		
14	10	6	3	600.00		
15	11	6	4	500.00		
16	12	3	1	400.00		
17	13	4	2	300.00		
18	14	6	2	200.00		
19	15	6	2	100.00		
20	16	3	6	-		

FIGURE 8.72   Excel input data and results.

```
Sub Win_3()
    Range("A1.E104").ClearContents        'Clean all work area
                            'Write titles to identify values
    Cells(1, "A") = "Initial capital"
    Cells(1, "B") = "Amount placed in each bet"
    Cells(1, "C") = "Prize factor"
    Cells(1, "D") = "How many repetitions for the experiment?"
    Cells(4, "A") = "Bet"
    Cells(4, "B") = "Die 1"
    Cells(4, "C") = "Die 2"
    Cells(4, "D") = "Credit available"
                            'Enter data into the program and display its values
    IC = Val(InputBox("How Much is your Initial Capital?",, 1000))
    Range("A2").Value = IC
    B = Val(InputBox("Amount placed in each bet?",, 100))
    Range("B2").Value = B
    F = Val(InputBox("Prize factor?",, 3))
    Range("C2").Value = F
    R = Val(InputBox("How many repetitions for this experiment?",, 10))
    Range("D2").Value = R
    Tot_Bet = 0             'Total number of bets until zero credit availability
    For Rep = 1 To R        'Execute R Repetitions
        EndYes = 0
        Bet = 0             'Initialize Bets counter for repetition
        C = IC              'Keep original initial capital
        Range("A5:D1004").ClearContents    'Clean repeat values
    While EndYes = 0
        Bet = Bet + 1                   'Increase the replay bet counter
        Cells(Bet + 4, "B") = "=RandBetween(1, 6)"     'Roll first die
        D1 = Cells(Bet + 4, "B")            'D1 takes die one value
        Cells(Bet + 4, "C") = "=RandBetween(1, 6)"
        D2 = Cells(Bet + 4, "C")            'Record in D2 second die value
        Cells(Bet + 4, "A") = Bet
        Cells(Bet + 4, "B") = D1            'Write in column B given value D1
        Cells(Bet + 4, "C") = D2
        If D1 = D2 Then
            C = C - B + (F * B)    'Dice equal. Deduct value bet and collect prize
        Else
            C = C - B                    'Dice different. Deduct value bet
        End If
        If C <= 0 Then
            EndYes = 1                  'If bettor goes bankrupt end While
        End If
        Cells(Bet + 4, "D") = C
    Wend
        Tot_Bet = Tot_Bet + Bet         'Accumulate all bets from all replays
```

```
    Next Rep                          'Run the next iteration
    Range("F1").Value = "Amount of bets until all the capital is lost"
    Cells(2, "F") = Tot_Bet / R    'Estimate the expected number of bets to ruin
End Sub
```

The program allows you to include different values as input data.

Example 40: Goal Bettor

There is an old method for playing flips, which consists of doubling the amount wagered each time you lose. For example, if you bet $X and lose, then bet $2X; if you lose again this time, then bet $4X and so on. However, if by following this policy it happens that the doubled bet is greater than the amount you have available, then you bet the remainder. Conversely, each time you win, the bet will be $X. Suppose the initial amount available is $5,000 the initial bet is $1,000 the prize is equal to the amount wagered, the probability of winning in a flip is 0.5 and the bettor's objective is to retire when he achieves $10,000. Simulate and estimate: (A) What is the probability of achieving the desired withdrawal amount? (B) How many flips are expected to be made until the bettor loses all his capital?

Solution

A macro program named L_Flips() will be used. The procedure for using Excel macros is detailed in Sections 2.14 to 2.18. During its execution, the program requests the player's initial capital (IC), the initial value to bet (B), the target amount to be reached (G), and the number of repetitions to perform (R). Although it will be solved for the example indicated input data, the program is flexible to receive different input values so that it allows to define diverse scenarios.

Ranges I2:I10001 and J2:J10001 contain the duration of each repetition or the number of bets placed on each replay (Figure 8.73). If you count the number of values present in those columns, you get the number of iterations executed to finish by goal achievement or bankruptcy. Outside of the macro, to count the repetitions in which the player achieves his objective, in cell L5 the following expression could be directly included:

=COUNTIF(I2:I10001;">0")

	A	B	C	D	E	F	G	H	I	J	K	L	M
1	Initial Capital	Start Betted Amount	Goal to Reach	How Many Repetitions for the Experiment				Repetition	Bets until the Goal is Reached	Number of bets until the player goes Bankrupt		Total Bets until the Goal is reached	Total Bets until Ruin
2	5,000	1,000	10,000	10,000				1		3		45,186	34,767
3								2	9				
4	Bet	Initial Capital	Amount Betted	Random Flip	Win?	End Capital		3	10			Times the goal was achieved	Times the player was ruined
5	1	5,000	1,000	0.633	N	4,000		4	6			5,021	4,979
6	2	4,000	2,000	0.761	N	2,000		5	7				
7	3	2,000	2,000	0.991	N	-		6	8			Answers A & B	
8								7		9		Probability of Reaching the Goal	Bets Until Bettor Loses All Capital
9								8		6		0.5021	7.0
10								9	12			Click to	
11								10	9			Throw the	
12								11	18			Coin	

FIGURE 8.73   Sample of the replica 10,000 and results.

However, in the macro was provided the following equivalent code:

```
Cells(5, "L") = Range("L5").Select
ActiveCell.FormulaR1C1 = "=COUNTIF(R[-3]C[-3]:R[9995]C[-3],"">0""")"
```

It was simulated for the input data shown in Figure 8.73 and a simulation length of 10,000 replicates.

The probability of achieving the desired goal is almost similar to that of going bankrupt: Around 0.5. On average a player is expected to lose his capital in seven bets. Although the average number of wagers a bettor is expected to make until the end of their betting session is approximately 16.

The macro Excel programming code appears below:

```
Sub Flips()
    Randomize
    Range("A1.f104").ClearContents
    Range("H2.J10001").ClearContents
    Cells(1, "A") = "Initial Capital"
    Cells(1, "B") = "Start Betted Amount"
    Cells(1, "C") = "Goal to Reach"
    Cells(1, "D") = "How Many Repetitions for the Experiment"
    Cells(4, "A") = "Bet"
    Cells(4, "B") = "Initial Capital"
    Cells(4, "C") = "Amount Betted"
    Cells(4, "D") = "Random Flip"
    Cells(4, "E") = "Win?"
    Cells(4, "F") = "End Capital"
    IC = Val(InputBox("How Much is your Initial Capital?",, 5000))
    Range("A2").Value = IC
    B = Val(InputBox("How Much are You Willing to Bet?",, 1000))
    Range("B2").Value = B
    G = Val(InputBox("How Much is Your Goal",, 10000))
    Range("C2").Value = G
    R = Val(InputBox("How Many Repetitions for the Experiment?",, 10000))
    Range("D2").Value = R
    AB = B              'Amount to Bet
    Tot_Bank = 0        'Total number of Bets until zero credit availability
    Tot_Goals = 0
    Cells(1, "H") = "Repetition"
    Cells(1, "I") = "Bets until the Goal is Reached"
    Cells(1, "J") = "Number of bets until the player goes Bankrupt"
    Cells(4, "L") = "Times the goal was achieved"
    Cells(4, "M") = "Times the player was ruined"
    For Rep = 1 To R          'Execute R Repetitions
        Goals = 0             'Amount of bets until you achieve your goal in
                              a repeat
        Bank = 0              'Amount of bets until it is ruined in a replay
```

```
EndYes = 0                    'While Control
BC = 0                        'Bet counter
Capi = IC
Range("A5:F104").ClearContents
While EndYes = 0                    'Repeat cycle until EndYes <>0
  BC = BC + 1                       'Increase bet counter
  Flip = Rnd()                      'Flip Coin
  Cells(BC + 4, "A") = BC    'Write the consecutive index of the bet
  Cells(BC + 4, "B") = Capi
  Cells(BC + 4, "C") = AB
  Cells(BC + 4, "D") = Flip

  If Flip <= 0.5 Then
    Cells(BC + 4, "E") = "Y"        'Bettor wins write Y
                                    'Show capital increase:
    Cells(BC + 4, "F") = Cells(BC + 4, "B") + Cells(BC + 4, "C")
  Else
    Cells(BC + 4, "E") = "N"        'If bettor loses write N
                                    'Show capital decrease
    Cells(BC + 4, "F") = Cells(BC + 4, "B") - Cells(BC + 4, "C")
  End If
  Win = Cells(BC + 4, "E")
  Capi = Cells(BC + 4, "F")
  If Win = "Y" Then
    AB = B
  Else
    If Capi >= 2 * AB Then
      AB = 2 * AB                   'Double the bet value
    Else
      AB = Capi                     'Bet the remaining capital
    End If
  End If
  If Capi >= G Or Capi <= 0 Then 'Decide if capital exceeds target or is
                                    depleted
      EndYes = 1
  End If
Wend
Cells(Rep + 1, "H") = Rep
If Capi >= G Then                           'If the capital reaches the goal
    Goals = Goals + 1
    Bet_Goal = BC
    Tot_Goal = Tot_Goal + BC
    Cells(Rep + 1, "I") = Bet_Goal    'Write number of goals achieved
                                       in a repetition
End If
If Capi <= 0 Then    'If the capital runs out
    Bank = Bank + 1    'Increase bankruptcies by 1
```

```
                Bet_Bank = BC
                Tot_Bank = Tot_Bank + BC
                Cells(Rep + 1, "J") = Bet_Bank    'Write Bankrupts achieved in the
                                                   repetition
      End If
      AB = B                                      'Reallocate the initial amount to bet
      Next Rep
      Cells(1, "L") = "Total Bets until the Goal is reached"
      Cells(1, "M") = "Total Bets until Ruin"
      Cells(2, "L") = Tot_Goal
      Cells(2, "M") = Tot_Bank
                'to count the repetitions in which the player achieves his objective
      Cells(5, "L") = Range("L5").Select
      ActiveCell.FormulaR1C1        '=        ""=COUNTIF(R[-3]C[-3]:
                                                  R[9995]C[-3],"">0"")"

      Cells(5,"M")=R-Cells(5,"L") 'Deduct from R times the bettor achieves his goal
      Loose_B = Cells(5,"M")        'Loose_B stores the number of times the bettor
                                    busts
      Cells(7, "L") = "Answers A & B"
      Cells(8, "L") = "Probability of Reaching the Goal"
      Cells(9, "L") = Cells(5, "L") / R
      Cells(8, "M") = "Bets Until Bettor Loses All Capital"
      Cells(9, "M") = Tot_Bank / Loose_B
'End Sub
```

Example 41: The Wheel of Fortune
A wheel of fortune has 20 equally spaced portions. Each equally spaced area contains a prize. Its distribution is as follows:

Prize $:	0	100	200	300	400	600	800	900	1,200	2,000
Slots:	4	3	2	2	2	3	1	1	1	1

The gambler has a $1,000 and the amount of a single bet is $500. The player will place an indeterminate round of bets as long as he has enough capital to cover the value of the bet. Simulate a length of 10,000 rounds to estimate: (A) Number of bets he expects to make until he has no money to continue. (B) Probability of winning a bet.

Solution
For a better approximation, the simulation length can be determined. Here an Excel applet or macro will be used to simulate 1,000 repetitions. The theoretical probability of winning on any bet can be calculated analytically and used to compare with the experimental result. This probability is obtained by adding those probabilities whose remuneration exceeds the amount bet. It is 0.35. Figure 8.74 shows the results of the

	A	B	C	D	E	F	G	H	I	J	K	L	M	N
1	Initial Balance:	1,000.00		Prize $:	0	100	200	300	400	600	800	900	1,200	2,000
2	Bet:	500.00		Wedges:	4	3	2	2	2	3	1	1	1	1
3				Probability:	0.20	0.15	0.10	0.10	0.10	0.15	0.05	0.05	0.05	0.05
4			Accumulated:	0	0.20	0.35	0.45	0.55	0.65	0.80	0.85	0.90	0.95	1
5									Results:					
6	Spin	Balance Before Bet	Balance minus Bet	Random Number	Price $	Available Balance		Average Number of Bets Placed Until the Capital is Exhausted		Probability of Winning a Bet				
7	1	1,000.00	500.00	0.302	100.00	600.00		12.3		0.349				
8	2	600.00	100.00	0.689	600.00	700.00								
9	3	700.00	200.00	0.712	600.00	800.00								
10	4	800.00	300.00	0.780	600.00	900.00		Click to Spin the Wheel of Fortune						
11	5	900.00	400.00	0.109	-	400.00								
12		400.00												

FIGURE 8.74 Input and results.

experiment executed using the macro Wheel_F. It is previously required to include in the sheet area those fixed texts that identify the input and output values.

The results indicate that according to the conditions of the bet, a bettor expects to make between 12 and 13 bets until withdrawing. Likewise, as expected, the probability of winning a bet is estimated to be close to 0.349. See Figure 8.74, cell K7.

The programming code for the Excel macro named Wheel appears below:

```
Sub Wheel_F()
  Randomize
  IC = Range("B1").Value          'Save in IC the amount available to play
  Bet = Range("B2").Value         'Save the fixed value to bet
  Tot_Wins = 0
  Tot_Bets = 0
  For Rep = 1 To 10000            'Repeat 10,000 times
    Range("A8.F1000").ClearContents
    Spin = 1                      'Spin Records the amount of bets in each
                                   repetition
    AC = IC                       'AC keeps initial amount available for the
                                   player to place their bets
    Cells(Spin + 6, "B") = IC     'Write IC on screen
    Wins = 0                      'Wins records the times you win in each
                                   repetition
    Endyes = 0                    'Control variable of each repetition
    While Endyes = 0              'As long as there are funds to bet on this round
      AC2 = AC
      Cells(Spin + 6, "A") = Spin       'Write the bet counter Spin on the screen
      Cells(Spin + 6, "C") = AC - Bet   'Deduct the amount betted
      ACAC = Cells(Spin + 6, "C")
      RR = Rnd()                  'Generates a random for the spin result
      Cells(Spin + 6, "D") = RR
```

```
                        'Then look for the RR for the result of the roulette spin
Cells(Spin + 6, "E") = "=LOOKUP(RC[-1],R4C4:R4C15,R1C5:R1C15)"
WR = Cells(Spin + 6, "E")      'Save the prize achieved in WR variable
AC = ACAC + WR                 'Update the available capital
Cells(Spin + 6, "F") = AC
If AC > AC2 Then               'If Balance Before Bet > Available
                                Balance, increase Wins

    Wins = Wins + 1             'Increase Wins by one
End If
If AC < Bet Then               'Check if there is capital left to
                                continue betting

    Tot_Bets = Tot_Bets + Spin 'Total bets made
    Endyes = 1                 'End While command
End If
Spin = Spin + 1                'Make another bet
Cells(Spin + 6, "B") = AC      'Write Available Capital
Wend
Tot_Wins = Tot_Wins + Wins     'Total times you win in a betting round
Next Rep
Cells(7, "I") = Tot_Bets / 10000   'Estimate the amount of bets until
                                    losing the capital
Cells(7, "K") = Tot_Wins / Tot_Bets   'Estimate the probability of winning
                                       at roulette
End Sub
```

Example 42: Gym Locks

A gym has lockers (closets) for people to store their clothes and other objects. Each user decides whether or not to place the only available padlock model. Suppose that:

- The value of the objects contained in any closet is distributed according to a normal with an average of $1,000 and a deviation of 200.
- If it does not have a lock, the chance that a thief will steal the items from the closet is 0.30; and 0.07 otherwise.
- In addition, with a probability of 0.20 at any time the user can lose his key, and resort to a locksmith to dismantle the old padlock and place a new one at a cost of $30. A new lock costs $50.

Simulate 10,000 closets to estimate the expected cost to a user for locking or not locking their closet.

Solution

In Figure 8.75, the range A1:H3 contains the input data corresponding to the cumulative theft risk probabilities. Cells O1:O2, although for one run, approximate responses to the expected costs of a user whether or not he uses a lock. Cell O1 using the expression =AVERAGE(N6:N10002), and O2 with =AVERAGE(O6:O10002).

	A	B	C	D	E	F	G	H	I	J	K	L	M	N	O
1							Y	N			Average Cost With Padlock ($):				68.4
2	Cumulative probability of theft without a lock:					0	0.3	1			Average Cost Without Lock ($):				181.7
3	Cumulative probability of theft with padlock:					0	0.07	1							
4															
5	Locker	Items Value ($)	Rand (Padlock)	Does it have a Lock?	Rand (Theft)	Did they Steal?	Rand (loss of key)	Lost Key?		Loss Due to theft ($)	Lock Cost ($)	Total Cost ($)		Cost With Lock	Cost Without Lock
6	1	976	0.918	N	0.945	N	0.609	N		0	50	50		0	50
7	2	807	0.253	Y	0.316	N	0.331	N		0	50	50		50	0
8	3	663	0.923	N	0.184	Y	0.487	N		663	50	713		0	713
9	4	509	0.359	Y	0.038	Y	0.892	N		509	50	559		559	0
10	5	915	0.860	N	0.253	Y	0.139	Y		915	130	1045		0	1045
11	6	962	0.339	Y	0.213	N	0.086	Y		0	130	130		130	0

FIGURE 8.75 Sample of first six lockers.

The expressions arranged in the cells of the row range B6:O6 are shown in the following table:

Cell	Content
A6	In toolbar HOME / Fill / Series … / Columns / Linear / Stop value 10000 / OK
B6	=INT(NORM.INV(RAND();1000;200))
C6	=RAND()
D6	=IF(C6<0.5;"Y";"N")
E6	=RAND()
F6	=IF(D6="Y";LOOKUP(E6;$F$3:$H$3;$G$1:$H$1);LOOKUP(E6;$F$2:$H$2; $G$1:$H$1))
G6	=RAND()
H6	=IF(G6<0.2;"Y";"N")
J6	=IF(F6="Y";B6;0)
K6	=IF(H6="Y";130;50)
L6	=J6+K6
N6	=IF(D6="Y";L6;0)
O6	=IF(D6="N";L6;0)

All expressions arranged in row 6, range A6:O6, are copied to range A7:O10005. After executing 20 runs, the expected cost with a lock is estimated to be $67.8; and $183 without use of a lock.

Example 43: Four Factories

A company manages four factories that produce the same item. All the production of the four factories is accumulated in a central warehouse. The inventory of the deposited product has the following distribution of origin and defects:

Factory of Origin	A	B	C	D
Percentage (%)	20	40	25	15
Defect rate (%)	1	2	1.3	0.5

A product is chosen at random and it turns out to be defective. (A) Determine the theoretical probability of once selected a defective item, this comes from factory B. (B) For a confidence level of 99% and an error of 0.5%, determine the length of the simulation and estimate the probability that a selected item comes from factory B given that it is a defective item.

Solution

A. According to Bayes' theorem, the theoretical probability of selecting a defective item which comes from factory B, is obtained by the following relationship:

$$P(B/D) = \frac{P(D/B) * P(B)}{P(D)} = \frac{0.02 * 0.40}{0.014} = 0.5714$$

Where the probability of selecting a defective item is obtained according to:

P(D) = P(D/A) + P(D/B) + P(D/C) + P(D/D) =
0.01 ∗ 0.2 + 0.02 ∗ 0.40 + 0.013 ∗ 0.25 + 0.005 ∗ 15 = 0.014

In summary, the theoretical probability that once a defective item is selected, it comes from a certain factory X is:

X	A	B	C	D
Theoretical probability $P(X/D) =$	0.1429	**0.5714**	0.2321	0.05377

B. To determine the simulation length, remember that the value to be estimated is a probability, so the following relationship is used:

$$n \geq \frac{z_{1-\alpha/2}^2}{4 * \in^2} = \frac{NORM.\ S.\ INV\ (0.005)}{4 * 0.005^2} = \frac{2.576^2}{4 * 0.005^2} = 66{,}349$$

A sample length of 66,400 articles will be used (Figure 8.76).

	A	B	C	D	E	F	G	H	I	J	K	L	M	N	O
1		Defects (%)	1%	2%	1.3%	0.5%									
2		Factory	A	B	C	D									
3		Probability:	0.2	0.4	0.25	0.15									
4	Cumulative		0	0.2	0.6	0.85	1								
5															
6										Factory	A	B	C	D	Total
7					Factory Defective Items					Total Defectives	134	539	210	59	942
8	Product	Random for Factory	Made by	Random for defects	A	B	C	D		Probability (Factory/Def.)	0.142	0.572	0.223	0.063	
9	1	0.0083	C	0.1345	0	0	0	0							
10	2	0.6910	B	0.5898	0	0	0	0							
11	3	0.5019	B	0.2006	0	0	0	0							
12	4	0.3874	B	0.7947	0	0	0	0							
13	5	0.3037	C	0.5757	0	0	0	0							

FIGURE 8.76 Simulation of conditional probabilities. Range A9:H66408.

Estimate P(B/D)

The estimated value for the probability that an item that was found to be defective was manufactured by factory B, P(B/D), is close to 0.572. Figures 8.77, 8.78, and 8.79 show the expressions used to perform the simulation.

	A	B	C	D	E
1		Defects (%)	0.01	0.02	0.013
2		Factory	A	B	C
3		Probability:	0.2	0.4	0.25
4	Cumulative	0	0.2	=+C4+D3	=+D4+E3
5					
6					
7					Factory Defective Items
8	Product	Random for Factory	Made by	Random for defects	A
9	1	=RAND()	=LOOKUP(B10;$B$4:$F$4;$C$2:$F$2)	=RAND()	=IF(C9= "A"; IF(D9<$C$1;1; 0); 0)
10	=A9+1	=RAND()	=LOOKUP(B11;$B$4:$F$4;$C$2:$F$2)	=RAND()	=IF(C10= "A"; IF(D10<$C$1;1; 0); 0)

FIGURE 8.77 Main expressions used in the range A9:E66408.

	F	G	H
1	0.005		
2	D		
3	0.15		
4	=+E4+F3		
5			
6			
7			
8	B	C	D
9	=IF(C9= "B"; IF(D9<$D$1;1; 0); 0)	=IF(C9= "C"; IF(D9<$E$1;1; 0); 0)	=IF(C9= "D"; IF(D9<$F$1;1; 0); 0)
10	=IF(C10= "B"; IF(D10<$D$1;1; 0); 0)	=IF(C10= "C"; IF(D10<$E$1;1; 0); 0)	=IF(C10= "D"; IF(D10<$F$1;1; 0); 0)

FIGURE 8.78 Defects in factories B, C, and D. Range F9:H66408.

	J	K	L	M	N	O
1	Factory	A	B	C	D	Total
2	Total Defectives	=SUM(E9:E66408)	=SUM(F9:F66408)	=SUM(G9:G66408)	=SUM(H9:H66408)	=SUM(K2:N2)
3	Probability (Factory/Def.)	=+K2/$O$2	=+L2/$O$2	=+M2/$O$2	=+N2/$O$2	

FIGURE 8.79 Formulas to estimate conditional probabilities.

Example 44: Multinational Employees

Twenty percent of the employees of a multinational company are engineers and another 20% are economists. 75% of engineers hold a management position and 50% of economists also, while of employees who are not engineers or economists, only 20% hold management positions. For his morning show, an interviewer randomly chose a manager from the multinational. (A) What is the probability that the manager interviewed is an engineer by profession? (A Use a simulation of length 10,000 employees. (B) Construct a confidence interval for the estimated value.

Solution

	Engineers	Economists	Others
% of employees	20%	20%	60%
Cumulative probability	0.20	0.40	1.0

	Status	
Profession	Executive	Non-Directive
Engineers	75%	25%
Economists	50%	50%
Others	20%	80%

The range of cells A1:J5 in Figure 8.80 contains the input data ordered in cumulative probability distributions, arranged to randomly obtain the status of each employee.

To determine the professional status of employee 1, cell D8 contains the expression:

=LOOKUP(C8; $B$2:$E$2; $C$1:$E$1)

This formula is copied to cells D9 to D10007. Cell C8 generates a random number =RAND().

In cell F8, the professional status of the employee is evaluated and it is randomly determined if he has managerial status or not, registering for the employee one of the following hierarchy:

	A	B	C	D	E	F	G	H	I	J
1		Profession	Eng	Eco	O			Economist	EcoM	EcoNM
2	Cumulative Probability:	0	0.20	0.40	1		Cumulative Probability:	0	0.50	1
3										
4		Engineer	EngM	EngNM				Other	OM	ONM
5	Cumulative Probability:	0	0.75	1			Cumulative Probability:	0	0.20	1
6										
7		Employee	Random	Profession	Random	Hierarchy		Managers in 10 thousand Employees	Profession	
8		1	0.6902	O	0.4371	ONM		1513	Engineer	
9		2	0.1067	Eng	0.0691	EngM		972	Economist	
10		3	0.7265	O	0.2495	ONM		1165	Other	
11		4	0.1634	Eng	0.0125	EngM				
12		5	0.6443	O	0.4579	ONM		Prob_Est(Eng/M)=	0.4145	

FIGURE 8.80 Display of input data and results.

EngM	EngNM	EcoM	EcoNM	OM	ONM
Engineer/ Manager	Engineer/ Non-Manager	Economist/ Manager	Economist/ Non-Manager	Other/ Manager	Other/ Non-Manager

In Figure 8.80, the range F9:F10007 contains a copy of the following expression arranged in cell F8:

=IF(D8="Eng";LOOKUP(E8;$B$5:$D$5;$C$4:$D$4);IF(D8="Eco";LOOKUP (E8;$H$2:$J$2;$I$1:$J$1);LOOKUP(E8;$H$5:$J$5;$I$4:$J$4)))

The above expression can be reduced in size length by using range names.

Cells H8 through H10 contain expressions to count the number of managerial employees in each of the three possible profession options and allow probability estimation. For example, cell H8 has the expression: = COUNTIF($F$8:$F$10007; "EngM")

Cells H9 and H10 contain =COUNTIF($F$8:$F$10007;"EcoM") and =COUNTIF ($F$8:$F$10007;"OM"), respectively.

In cell I12, the Bayesian conditional probability is estimated using the expression =H8/(H8+H9+H10).

After 20 runs, the conditional probability of having selected a manager as the interviewee, who in the end turns out to be an engineer, is close to 0.4054.

Example 45: Guess a Whole Number
A game in a betting house consists of guessing a whole number chosen by the house between the values 0 to 100. The bettor has up to 10 attempts. Each time it fails, the house tells you if the most recent number tried is less than or greater than the secret

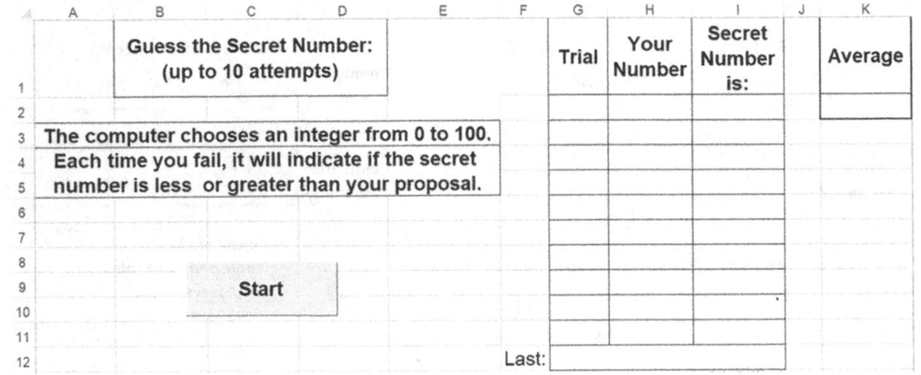

						Trial	Your Number	Secret Number is:	Average
1		Guess the Secret Number: (up to 10 attempts)							
2									
3	The computer chooses an integer from 0 to 100.								
4	Each time you fail, it will indicate if the secret								
5	number is less or greater than your proposal.								
6									
7									
8									
9			Start							
10									
11									
12						Last:			

FIGURE 8.81   Preliminary screen for data entry.

number. Develop a simulation model to mimic this experiment and estimate the expected number of attempts to achieve the goal.

Solution

Figure 8.81 shows the preliminary layout of texts to be filled in to instruct the user about the game conditions and show its sequence of results. All texts must be arranged by the user. Pressing Start will run the applet called Guess().

The average number of attempts to guess the random value generated by the machine will be estimated using the following subprogram Guess():

```
Sub Guess()
    Dim x As Byte, sn As Byte
    times = InputBox("How many tries?",, 2, 1, 1)
    For Rep = 1 To times
        Range("G2.I11").ClearContents
        x = Int(Rnd * 101)
        try = 1
        Total = 1
        OutRep = "N"
        Do While OutRep = "N"
            sn = InputBox("Guess: Enter an integer between 0 and 100" & vbCrLf _
              & "You can make up to 10 attempts", "Attempt number" & try,, 1, 1)
            If x <> sn Then
                If x < sn Then
                    zon = "Smaller"
                Else
                    zon = "Greater"
                End If
                Cells(1 + try, "G") = try
                Cells(1 + try, "H") = sn
```

```
      Cells(1 + try, "I") = zon
      try = try + 1
      Tot = Tot + 1
      If try > 10 Then
        MsgBox "Ten failed attempts" & vbCrLf & "The secret number is" & x
        OutRep = "Y"
        If OutRep = times Then
          Exit Sub
        End If
      End If
     End If
     If sn = x Then
        MsgBox "Very good!" & vbCrLf & "You guessed the secret number" &
  x & ", in" & try & "Attempts"
        Cells(12, "G") = x
        OutRep = "Y"
        Tot = Tot + 1
     End If
    Loop
   Next Rep
   Average = Tot / times
   Cells(2, "K") = Average
 End Sub
```

Section 2.18 of chapter 2 explains the corresponding procedure to follow for the automation of macro subprograms (Figure 8.82).

Once the Start command is pressed, the applet is told how many times the guessing game will be played. Then it records each attempt to guess the secret number. Every time you fail, it will tell you if the secret number is less than or greater than your

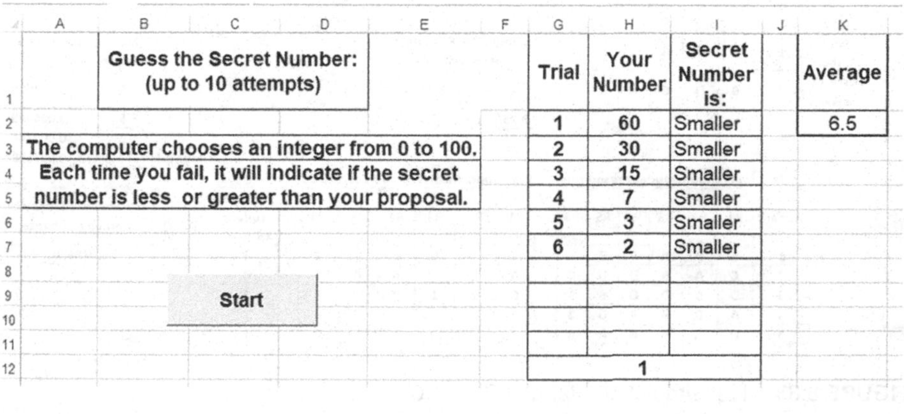

	A	B	C	D	E	F	G	H	I	J	K
1		Guess the Secret Number: (up to 10 attempts)					Trial	Your Number	Secret Number is:		Average
2							1	60	Smaller		6.5
3	The computer chooses an integer from 0 to 100.						2	30	Smaller		
4	Each time you fail, it will indicate if the secret						3	15	Smaller		
5	number is less or greater than your proposal.						4	7	Smaller		
6							5	3	Smaller		
7							6	2	Smaller		
8											
9			Start								
10											
11											
12							1				

FIGURE 8.82   Secret number simulator screen.

guess. Cell K2 shows the average number of attempts to guess the secret number. After 20 independent runs, it is estimated at 6.5 attempts.

Example 46: Questionnaire

A person answers a questionnaire that contains six questions. Each question offers four answer alternatives: A, B, C, or D. The evaluated person chooses at random with equal probability each of his answers. To pass, at least four correct answers are required. (A) Indicate the theoretical probability that the person passes. (B) Simulate for a simulation length of 10,000 people and estimate the probability of approval.

Solution

A. The evaluated person can choose with equal probability a response from among four possible values. The probability of choosing option A, B, C, or D is 0.25. Since you must get at least four out of six possible questions correct, the probability distribution is binomial:

$$P(Aciertos\ X \geq 4) = \sum_{x=4}^{6} \binom{6}{x} 0.25^x * 0.75^{6-x} = 1 - BINOM.DIST\,(3;\ 6;\ 0.25;\ 1)$$

$$P(Hits\ X \geq 4) = \sum_{x=4}^{6} \binom{6}{x} 0.25^x * 0.75^{6-x} = 1 - BINOM.DIST\,(3;\ 6;\ 0.25;\ 1)$$

Consequently, the probability of passing the questionnaire for any individual who responds at random is 0.037598.

B. Figure 8.83 the range of cells A1:G2, assumes a random sequence of correct answers to the six questions of the questionnaire. The sequence B, B, D, C, A, A corresponds to the pattern to be compared with the six answers given by those evaluated.

A	B	C	D	E	F	G	H	I	J	K	L	M	N	O	P	Q	R	S	T
Question Number	1	2	3	4	5	6			Cumulative Probability of Responding									Right Answers	Binomial Probability
Correct Answer	B	B	D	C	A	A			A	B	C	D						4	0.032959
								0	0.25	0.5	0.75	1						5	0.004395
																		6	0.000244
																		Theorical Probability:	0.037598
		Six random student responses							Evaluation: Incorrect = 0 / Correct = 1								Correct Student Answers	Simulated Probability to Approve by Chance	
	Student	R1	R2	R3	R4	R5	R6		R1	R2	R3	R4	R5	R6						
	1	B	B	A	B	B	A		1	1	0	0	0	1			3	0.038600	
	2	C	A	B	D	D	C		0	0	0	0	0	0			0		
	3	D	B	D	D	B	B		0	1	1	0	0	0			2		
	4	A	B	A	B	D	B		0	1	0	0	0	0			1		
	5	D	A	A	D	A	B		0	0	0	0	1	0			1	Activar Windows	

FIGURE 8.83 Test simulation. Range B8:B10007.

To mimic the random responses given by 10,000 individuals, the expression

=LOOKUP(RAND(); $I$3:$M$3; $J$2:$M$2)

It is arranged in the range of cells C8:H10007. This relationship randomly chooses six answers for each individual, which are compared one by one, in the range J8:O10007, displaying a 1 if the answer is correct or a 0 otherwise. For example, in cell C8, IF(C8= $B$2; 1; 0) evaluates the answer of the first evaluated to the first question. If they match, fill in 1.

In column Q, the range of cells Q8:Q10007 executes the sum of the ones or zeros in each row of the range J8:O10007, which corresponds to counting the number of hits for each evaluated person. For example, =SUM(J8:O8) gets the score of the first test taker.

Finally, cell S8 estimates the probability of passing the test by using the expression:

=COUNTIF(Q8:Q10007; ">3") / 10000

Cell S8 presents the estimated probability that a student randomly passes the test. The estimated value in the run is 0.00386. If the results obtained by about 20 simulations are run and averaged, a more precise estimate is achieved.

Cells B1 to G2 contain the pattern of correct answers for the test. It can be filled with any combination of letters.

Cells I1 to M3 directly receive the cumulative probability values. To generate the six random answers given by a student, all cells in the range C8:H10007 contain the expression:

=LOOKUP(RAND();$I$3:$M$3;$J$2:$M$2)

In the cell range J8:O10007, the six answers given by the students are compared one by one against the correct pattern. For each of the six responses per student, expressions such as the following are available in cells range J8:O10007.

=IF(C8=$B$2;1;0) =IF(D8=$C$2;1;0) =IF(E8=$D$2;1;0)

=IF(F8=$E$2;1;0) =IF(G8=$F$2;1;0)

Figure 8.84 shows the complementary expressions arranged to complete the simulation.

Example 47: Hostesses and Airplanes
One airline offers 20 daily morning flights. Each flight is attended by two flight attendants. Additionally, the company has a reserve (ad hoc) group made up of eight hostesses available, who are available to replace in case any of the permanent hostesses are missing. The probability distribution of the number of flight attendants missing on any given day is as follows:

Stewardesses	0	1	2	3	4	5	6	7	8	9
Probability	0.03	0.05	0.08	0.10	0.10	0.15	0.20	0.15	0.10	0.04

	O	P	Q	R	S	T
1					Right Answers	Binomial Probability
2					4	=BINOM.DIST(S2;6;0.25;0)
3					5	=BINOM.DIST(S3;6;0.25;0)
4					6	=BINOM.DIST(S4;6;0.25;0)
5					Theorical Probability:	=SUM(T2:T4)
6						
7	R6		Correct Student Answers		Simulated Probability to Approve by Chance	
8	=IF(H8=$G$2;1;0)		=SUM(J8:O8)		=COUNTIF(Q8:Q10007;">3")/10000	
9	=IF(H9=$G$2;1;0)		=SUM(J9:O9)			
10	=IF(H10=$G$2;1;0)		=SUM(J10:O10)			

FIGURE 8.84   Additional formulas for calculating probability.

Strict regulations do not allow a plane to fly without both flight attendants and force its cancellation. The cost of using a reserve hostess is $1,000. The average income for each plane is normal with a mean of $400,000, and a deviation of $50,000. If a flight is canceled, a fixed loss of $200,000 is incurred. The cost of flying a plane is $250,000. To estimate the average daily profit, determine the length of the simulation, with 95% confidence and an error of 3%. Simulate and estimate: (A) Fraction of utilization of reserve hostesses. (B) Probability of canceling a flight due to not having stewards. (C) Average daily profit.

Solution
In Figure 8.85, range H6:J9, the sample size or simulation length, is determined. It starts with a 1,000-day simulation. For each day the number of missing hostesses is determined. The range A1:K3 contains, as input data, the probability distribution of absent hostesses on any given day. Cells in the range A5:B11 receive the input values. The simulation is deployed in the range A15:J1014.
The formulas arranged in the range A15:J1014 are shown in the following table:

Cell	Content
A15	On the toolbar: HOME / Fill / Series … / Columns / Linear / Stop value 1000 / OK
B15	=RAND()
C15	=LOOKUP(B15;$A$3:$K$3;$B$1:$K$1)
D15	=IF(C15>$B$9;C15-$B$9;0)
E15	=IF(C15>$B$9;$B$9;C15)
F15	=INT((20-D15)*($B$5+$B$6*(NORMSINV(RAND()))))
G15	=$B$11*(20-D15)
H15	=D15*$B$7
I15	=$B$10*E15
J15	=F15-G15-H15-I15

	A	B	C	D	E	F	G	H	I	J	K
1	Missing Hostesses:	0	1	2	3	4	5	6	7	8	9
2	Probability:	0.03	0.05	0.08	0.10	0.10	0.15	0.10	0.15	0.14	0.10
3		0	0.03	0.08	0.16	0.26	0.36	0.51	0.61	0.76	0.9	1
4											
5	Entry/Flight ($):	300,000.00		A) % Use Hostess Reservation:							
6	Deviation ($):	25,000.00		0.6225				z(0.975):	1.9600	3.841459	
7	Loss for Cancel a Flight($):	200,000.00						Variance:	2.8107E+11		
8				B) P(Of Canceling a Flight):				Error 3%:	56,606.15	3204256341	
9	Booking Hostesses:	7		0.01715				n >	337		
10	Cost of Hostess Reservation ($):	1,000.00									
11	Plane Cost/flight ($):	200,000.00		C) E(Daily Expected Utility)							
12				1,886,871.70							
13											
14	Day	Rand	Hostesses Missing	Flights Cancelled	Booking Hostesses Used	Income for Flights Made	Cost for Aircraft Flown	Loss for Cancelled Flights	Cost for Booking Hostesses	Daily Utility	
15	1	0.865	8	1	7	6,406,061.00	3,800,000.00	200,000.00	7,000.00	2,399,061.00	
16	2	0.747	7	0	7	5,126,948.00	4,000,000.00	-	7,000.00	1,119,948.00	
17	3	0.154	2	0	2	6,505,494.00	4,000,000.00	-	2,000.00	2,503,494.00	
18	4	0.016	0	0	0	6,003,433.00	4,000,000.00	-	-	2,003,433.00	

FIGURE 8.85 Sample flight simulation of 1,000 days.

The simulation length is determined for the daily utility variable displayed in column J. From the toolbar: **DATA / Data Analysis / Histogram / Input Range J15:J1014 / Chart Output** allow you to observe a Normal distribution for the utility. In addition, **Descriptive Statistics / Summary Statistics** shows similar values for Mean, Median, and Mode. Consequently, the sample size method based on the Central Limit Theorem will be used.

We will use 1,000 days sample size. For a significance level of 95%, $\alpha = 0.05$, and $Z(1 - \alpha/2) = Z(0.975) = $ NORM.S.INV(0.975) = 1.96. From a 1,000-pilot sample for the daily utility, we estimate an average utility in cell E12. It is \$1,886,871.70. Its 3% error, cell H8, is \$56,606.15.

Cell ranges G6:I9 have formulas to get a sample size for the simulation length.

Cell	Content
H6	=NORM.S.INV(0.975)
I6	=H6*H6
H7	=VAR(J15:J1014)
H8	= D12*0.03
I8	=H8*H8
I9	=ROUNDUP((I6*H7)/I8;0)
D12	=AVERAGE(J15:J1014)

After calculations, a size sample (337) smaller than the pilot sample of 1,000 days is obtained. After running 20 simulations, each of a 1,000 days, and 20 flights/day, the following results are estimated: (A) Fraction of utilization of reserve hostesses: 62.4%. (B) Probability of canceling a flight due to not having stewards: 0.0173. (C) Average daily profit: 1,891,646.5, cell D12.

Example 48: Walking Drunk
Suppose a drunk starts a walk on the corner of a block. Assume that the probability of going north, south, east, or west is the same each time he reaches another corner. If the alcoholic walks 15 blocks, simulate 10,000 drunken binges and determine: Probability that he will end at a distance less than or equal to two blocks from the point where he started.

Solution
The input values corresponding to the cardinal direction options and their probabilities are entered in the cell range A1:E2 (Figure 8.86). The drunkard starts at a corner, represented

	A	B	C	D	E	F	G	H	I	J	K
1	Dirección:	North	South	East	West				Distance from Origin		Distance in last Drunkenness (blocks):
2	0	0.25	0.5	0.75	1				5		5
3											
4	Block	Initial Coordinate	Initial Coordinate	Rand	Cardinal Direction	Final Coordinate	Final Coordinate				How many times was the drunk two or less blocks away?
5	1	0	0	0.801	West	-1	0				0
6	2	-1	0	0.251	South	-1	-1				
7	3	-1	-1	0.863	West	-2	-1				Probability:
8	4	-2	-1	0.455	South	-2	-2				0
9	5	-2	-2	0.953	West	-3	-2				
10	6	-3	-2	0.938	West	-4	-2				
11	7	-4	-2	0.563	East	-3	-2				
12	8	-3	-2	0.493	South	-3	-3				
13	9	-3	-3	0.522	East	-2	-3				
14	10	-2	-3	0.820	West	-3	-3				Click to Simulate a Walking
15	11	-3	-3	0.693	East	-2	-3				
16	12	-2	-3	0.259	South	-2	-4				
17	13	-2	-4	0.676	East	-1	-4				
18	14	-1	-4	0.454	South	-1	-5				
19	15	-1	-5	0.547	East	0	-5				

FIGURE 8.86 Simulation of the journey of a drunk.

on the Cartesian coordinate axes with the point (0; 0). A uniform random number between 0 and 1 allows you to model the direction he follows when walking a block.

The cell range A5:G19 contains instructions for simulating a 15-block trajectory. Their Excel expressions are listed in the following table:

Cell	Content
B5	0
C5	0
B6	=F5
C6	=G5
D5	=RAND()
E5	=LOOKUP(D5; $A$2:$E$2; $B$1:$E$1)
F5	=IF(OR(E5="North"; E5="South"); B5; IF(E5="East"; B5+1; B5-1))
G5	=IF(OR(E5="East"; E5="West"); C5; IF(E5="North"; C5+1; C5-1))

The range D5:G5 can be copied into the range D6:G19. Similarly, the B6:C6 range over the B7:C19 range. The macro named Drink() is used to make thousands of trips of 15 blocks each. The applet writes each distance calculated from cell K2 along column I.

```
Sub Drink()
    n = Val(InputBox("How many runs? = ", "Drinkers", 10))
    For i = 1 To n
        Cells(i + 1, "I") = Range("K2").Value
    Next i
End Sub
```

Cell K2, =ABS(F19)+ABS(G19, records the distance at which the drunk remains at the end of each walk. This value is passed to the macro and from there it is recorded in column I.

In cell K5, the expression =COUNTIF(I2:I50001;"<=2") counts the times that the drunk is two or less blocks from the origin.

Finally, for a simulation sample of 50,000 drunkenness, cell K8 = K5/50000 will show the probability that the drunk is two blocks or less from the origin of his trajectory. This value is approximately 0.155.

Example 49: The Thief of Baghdad

The thief of Baghdad is imprisoned in a dungeon that has three doors: one leads to freedom in just one day, another to a long tunnel, and the third to a short tunnel. The tunnels have no exit and the thief must return to his cell. When the thief returns he will

	A	B	C	D	E	F	
1		Free	Long_T	Short_T			
2		0.15	0.4	0.45			
3	0	0.15	0.55	1			
4							
5							
6	Attempt	Rand	Selected Door	Time (days)		Thief's Expected Time to Escape	
7	1	0.499	Long_T	2		5.4357	days
8	2	0.351	Long_T	2			
9	3	0.179	Long_T	18			
10	4	0.722	Short_T	2			
11	5	0.770	Short_T	2			

FIGURE 8.87 Simulation of the escape of the Baghdad thief.

attempt his next escape. However, his past experience does not help his next choice of door to choose because he lacks memory. That is, a very forgetful Markovian thief is assumed. On any given trial, the probability that the thief chooses the door to freedom is 0.15, the long tunnel is 0.40, and the short tunnel is 0.45. The travel time for any tunnel option is never less than two days, although for the long tunnel it is exponential with an average of nine days, and for the short tunnel it is exponential with an average of 3.5 days. Simulate 10,000 attempts and determine the thief's expected time to escape.

Solution
Input data is entered into Figure 8.87, cell range A1:D3. The range A7:D10006 provides the values to perform the simulation.

The following table contains the expressions available to carry out the simulation:

Cell	Content
A7	On the toolbar: HOME / Fill / Series ... / Columns / Linear / Stop value 10000 / OK
B7	=Rand()
C7	=LOOKUP(B7;$A$3:$D$3;$B$1:$D$1)
D7	=IF(C7="Free";1; IF(C7="Long_T"; MAX(2; INT(-9*LN(RAND())))); =MAX(2;INT(-3.5*LN(RAND()))))))
F7	=AVERAGE(D7:D10006)

The contents of the range A7:D7 must be copied to the range A8:D10006. Once the simulation is run, the average number of days to escape is indicated in cell F7. After executing 10,000 escape attempts, during 20 repetitions, the escape time expected by the fugitive is estimated at 5.3 days.

Example 50: Randomness of Success: Luck Counts
Set up an experiment to quantify the evolution of the particular success of individuals subjected to unpredictable events that may affect them positively or negatively. At the

beginning each individual receives a capital of $C_i(0) = 500$ monetary units (MU); in addition he is provided with intelligence and skills contained in his individual talent T_i, a Normal random variable with mean 0.6 and standard deviation 0.1. The individual talent T_i grows by the acquired experience that comes with age, so that the individual improves his talent semester by semester, being able to accumulate over time up to an additional 10% of the talent initially assigned. During each of the 80 semesters each person is subjected to a chance event with probability 0.5 that may or may not be adverse. If the subject is favored by a good luck event his capital $C_i(t)$ is increased according to the following random multiplicative factor:

Event Factor	1	2	3	4
Probability	0.35	0.30	0.25	0.10

In case an unlucky event happens to him his capital is decreased according to the random quotient factor in the table above. If a subject arrives at a period totally undercapitalized he continues this way until the end. For a confidence level of 99% and a maximum error of 1%, determine the length of the simulation to estimate: (A) the probability that an initially talented individual (Ti >= 0.8) ends up being a millionaire (Wealth > 10^6). (B) Assuming the model is valid, the number of millionaires expected to be in the world at the start of 2023.

Solution
To determine the length of the experiment, it is necessary to identify the type of the variable of interest. In this case, it is a matter of estimating a probability, proportion or fraction. Consequently, the corresponding formula will be used:

$$n \geq \frac{Z_{\alpha/2}^2}{4 * \varepsilon^2}$$

(a) $1-\alpha = 0.99$, (b) $\alpha= 0.01$, (c) $\varepsilon = 0.01$, and (d) $Z_{\alpha/2} = Z_{0.005} = 2.5758$
  The Excel sheet gives the standard normal value by: =NORM.S.INV(0.005)

$$n \geq \frac{Z_{0.005}^2}{4 * 0.01^2} = \frac{6.6348966}{0.004}$$

$n \geq 16{,}588$ individuals

That is, with 1% precision and 99% confidence, to estimate the probability that a talented individual becomes a millionaire, it is necessary to include at least 16,588 individuals in the simulation experiment.
  In the spreadsheet, 17,000 individuals are digitally located, each provided with their respective set of personal characteristics that represent their intelligence, abilities, skills, and knowledge called talent T_i, randomly generated from an assumed Normal random variable, with an average of 0.6 and a standard deviation of 0.1: $T_i \sim$ Normal(0.6; 0.1) (Figure 8.88).

FIGURE 8.88   Individuals talent.

At the beginning, each individual receives a capital of $C_i(0)$ = 500 UM. During 80 semi-annual periods equivalent to 40 years, each observed person is exposed to a chance event with probability 0.5 that could be favorable or adverse. Every six months, if the subject is favored by a good luck event his capital $C_i(t)$ will increase according to the event factor F.

$$C_i(t) = F * C_i(t - 1) \text{ (MU) if } R < T_i * (1 + t * 0.0125)$$

Where i corresponds to the individual number i = 1, 2, 3, …, 17,000 and t represents the semester period t = 1, 2, 3, …, 80

If an unfortunate event affects an individual in period t, his capital is reduced or divided according to the event factor F:

$$C_i(t) = C_i(t - 1))/F \text{(MU) if } R < T_i * (1 + t * 0.0125)$$

In Figure 8.89, cells A2:H5 contain the input values. Cell A4 receives the initial capital to be delivered to each individual. Cell B4 contains the probability of a favorable event occurring.

	A	B	C	D	E	F	G	H	I	J	K	L	M	N
1														
2				Input Data							Experiment Responses			
3	Initial Capital per Individual (MU):	Probability (Good Luck)		Event Factor (F)	1	2	3	4		Millionaires (Wealth >= 10^6 MU)	Maximum Individual Wealth (MU):	Individual Talent		Bankruptcy (Final Wealth < 1 MU)
4	500.00	0.5		Probability	0.35	0.30	0.25	0.10		137	544,195,584,000.00	Maximum	Minimum	12,788
5					0	0.35	0.65	0.90	1.0			0.98	0.17	
6														
7														
8				1	2	3	4	5	6	7	8	9	10	11	12
9	Individual	Initial Talent	Capital Period 1	Cap Period 2	Cap Period 3	Cap Period 4	Cap Period 5	Cap Period 6	Cap Period 7	Cap Period 8	Cap Period 9	Cap Period 10	Cap Period 11	Cap Period 12
10	1	0.72	500	125	42	42	42	42	42	14	56	28	28	28
11	2	0.56	250	1,000	1,000	1,000	1,000	500	500	1,500	3,000	1,000	250	250
12	3	0.66	250	250	250	250	63	63	31	8	2	1	1	1

FIGURE 8.89   Arrangement of inputs and some results.

The starting talent assigned to each individual is displayed in cell block B10:B17009 (Figure 8.89). This range must be filled with:

=MIN(ROUND(NORM.INV(RAND(); 0.6; 0.1); 2);1)

The MIN and ROUND functions guarantee a random talent no greater than 1 and rounded to two decimal places.

In case of good fortune, the impact of the event can double or even quadruple the estate. Although if the factor is equal to 1, it will not have consequences. This is simulated for the cell D10 according to:

=IF(C10>=0.5;IF(RAND()<$B$4;IF(RAND()<$B10*(**1+D$8*0.00125**);
    LOOKUP(RAND();$D$5:$H$5;$E$3:$H$3)*C10;C10);C10/ LOOKUP
    (RAND();$D$5:$H$5;$E$3:$H$3));0)

Cells D5:H5 collect the accumulated probabilities of the factor random variable, to determine the magnitude of the event that occurred. The cumulative distribution of probability to determine the magnitude of the event is defined in the range D5:H5 as described below:

D5 = 0 E5 = D5 + E4 F5 = E5 + F4 G5 = F5 + G4 H5 = G5 + H4

The value of the factor is obtained using the Lookup function. This function receives as arguments a random value and the cells to compare and find the value of the factor.

=LOOKUP(RAND(); $D$4:$H$4; $E$2:$H$2)

The contents of cell D10 must be copied to cell range E10:CD10, and then to cell range D11:CD17009. Cells C10 to C17009 do not consider an increase in talent by experience and excludes the expression highlighted in bold above. La expression C10>=0.5 checks if the individual owns at least 0.5 monetary units. Otherwise your capital remains at 0.

=IF(RAND()<$B$3 generates a random number and if it is less than 0.5 assumes a lucky event. However, a favorable event is randomized to the talent to determine its impact.

In cell J4 all the cells that correspond to the individual final capital after 40 years are verified from range CD10:CD17009 (Figure 8.89). The formula in cell J4 =COUNTIF(CD10:CD17009;">=1000000") counts individuals who achieve more than one million monetary units.

The result obtained in just one run was 137. See cell J4 in Figure 8.89. So the probability of being a millionaire corresponds to 137 out of 17,000. That is, the probability of being a millionaire is estimated at 0.0081. Multiplying this value probability by the world population gives the estimated number of millionaires. Consequently, the expected number of millionaires in a population of 8 billion individuals in 2023 is approximately 0.9%, equivalent to 64,470,588 humans. It is convenient to run additional simulations to average and improve the estimate.

Cell CK3, Figure 8.90, shows that 80.89% of individuals fall into extreme poverty; while a minority manages to accumulate most of the capital.

	CE	CF	CG	CH	CI	CJ	CK	CL
					Frequency Distribution of Individual Capital			
	Successful Talented (C>500,000 & T>0.8)	Successful Untalented (C>500,000 & T<0.6)	Average Capital for Talented Population		Richness Intervals (MU)	Individuals	%	Accumulated %
	23	14	5,078,639.11		10.00	13,752	80.89	80.89
					50.00	835	4.91	85.81
	Global Richness (MU)	Individual Average Net Worth (MU)	Number of Talented Millionaires		100.00	337	1.98	87.79
	615,184,994,638.14	36,187,352.63	24		1,000.00	836	4.92	92.71
					10,000.00	600	3.53	96.24
80					50,000.00	277	1.63	97.86
Financial Wealth at the end (MU)	Successful Talented? (C>500,000 & T>0.8) (0/1)	Successful Untalented? (C>500,000 & T<0.6) (0/1)	Final Wealth Amount of Talented Individual (T>0.8) (MU)	Tal & Mill?	100,000.00	64	0.38	98.24
0.00	0	0	-	0	500,000.00	131	0.77	99.01
7593.75	0	0	-	0	1,000,000.00	31	0.18	99.19
0.00	0	0	-	0	5,000,000.00	64	0.38	99.57
15.63	0	0	-	0	>5,000,000.00	73	0.43	100
0.00	0	0	-	0	Total:	17,000	Individuals	

FIGURE 8.90 Partial view of results.

To estimate the probability that an initially talented individual (Ti >= 0.8) ends up being a millionaire (Wealth > 106 UM), it is necessary to review the values of the CH column.

The range CH10:CH17009 from Figure 8.90 contains the expression:

=IF(AND(B10>=0.8;CD10>=10^6);1;0)

That verifies talent and capital conditions, placing 1 when the individual meets both.

The number of talented individuals and at the same time millionaires is obtained in cell CG6 by adding all ones appearing in the cell block CH10:CH17009 using the following expression:

=SUM(CH10:CH17009)

Cell CG6, see Figure 8.90, displays the value 24 individuals out of 17,000. Consequently, the probability that an individual is a millionaire and talented is estimated at 0.00142. That is, something less than 1.4 individual for every 1.000 people.

As previously indicated, to improve the robustness of the estimates, it is convenient to carry out about 20 runs. Figure 8.91 presents the result of 20 executions, whose responses, when averaged, offer better estimates.

The average number of millionaires after 20 executions is estimated at 147.4 out of 17,000. Then, the probability that an individual be a millionaire is 0.0086765. See cell B2 in Figure 8.91. Consequently, for a population of 8 billion, the number of millionaires in the year 2023 is estimated at 69,412,000 persons. This result satisfactorily resembles the pattern of world wealth.

The average number of individuals who, being talented, also become millionaires is 28.5, which corresponds to a probability of 0.0016765. See cell C2 in Figure 8.91.

In a new sheet, Figure 8.92 presents a partial view of the result of three independent runs. Each sample is ordered with respect to the final assets of each individual. It is observed how not necessarily the most talented individuals (CI) achieve the greatest fortune.

	Number of Millionaires (Wealth >= 10^6)	Number of Talented Millionaires		Year	World Population (Individuals)	%	Millionaires (Individuals)		
Average:	147.4	28.5		2021	7,912,666,000	0.00790	62,483,000 →Credit Suisses		
				2023	8,000,000,000	0.00867	69,341,176	2021	
							↓		
Run	Millionaires	Tal & Mill		Run	Millionaires	Tal & Mill	Simulated Results 2023		
1	151	29		11	157	36			
2	163	37		12	135	20			
3	161	31		13	162	31			
4	147	30		14	133	25			
5	131	31		15	156	26			
6	132	32		16	148	28			
7	161	39		17	158	28			
8	122	18		18	146	25			
9	157	36		19	143	24			
10	147	20		20	137	24			

FIGURE 8.91   Multiple runs.

Sample 1				Sample 2				Sample 3		
Individual	Initial Talent	Final Wealth (MU)		Individual	Initial Talent	Final Wealth (UM)		Individual	Initial Talent	Final Wealth (UM)
4474	0.68	18,874,368,000.00		147	0.85	60,466,176,000.00		4740	0.78	21,233,664,000.00
12675	0.72	17,915,904,000.00		486	0.86	11,943,936,000.00		16610	0.83	8,957,952,000.00
7466	0.84	6,718,464,000.00		13629	0.68	10,077,696,000.00		10579	0.75	3,981,312,000.00
9440	0.77	2,519,424,000.00		16489	0.91	8,957,952,000.00		11483	0.82	1,889,568,000.00
13255	0.67	2,519,424,000.00		14527	0.75	3,779,136,000.00		13331	0.68	1,769,472,000.00
12874	0.75	1,572,864,000.00		15148	0.81	2,239,488,000.00		1657	0.81	839,808,000.00
10527	0.81	1,492,992,000.00		13793	0.68	1,990,656,000.00		16660	0.69	663,552,000.00
13618	0.63	1,327,104,000.00		9224	0.69	1,889,568,000.00		2557	0.80	629,856,000.00
872	0.72	1,259,712,000.00		16645	0.75	1,327,104,000.00		9262	0.88	497,664,000.00
2730	0.63	1,259,712,000.00		14038	0.84	1,179,648,000.00		11520	0.76	393,216,000.00

FIGURE 8.92   Individual positioning of talent and fortune.

After executing numerous runs, it is verified that the frequency of accumulated individual capital is close to what is stipulated by the well-known Pareto principle: in the end, very few hoard large fortunes, while the majority are left with few monetary units. Each simulation shows that in the long run the extremely wealthy few will emerge while the vast majority will be plunged into financial misfortune. Furthermore, those who end up very rich are not necessarily the most competent. Likewise, the unpredictability of events can lead to individuals with different talents managing to accumulate similar amounts of capital.

Below are additional expressions and screenshots with formulas and data entered to perform this simulation experiment.

The capital accumulated by each individual, after the first semester, is declared in the range C10:C17009 (Figure 8.93) using the following expression:

=IF(RAND()<$B$4;IF(RAND()<$B10;LOOKUP(RAND();$D$5:$H$5;$E$3:$H$3)*$A$4; $A$4);$A$4/ LOOKUP(RAND();$D$5:$H$5;$E$3:$H$3))

	Input Data							Experiment Responses				
Initial Capital per Individual (MU):	Probability (Good Luck)	Event Factor (F)		1	2	3	4	Millionaires (Wealth >= 10^6 MU)	Maximum Individual Wealth (MU):	Individual Talent		Bankruptcy (Final Wealth < 1 MU)
500.00	0.5	Probability		0.35	0.30	0.25	0.10	148	14,155,776,000.00	Maximum	Minimum	12,855
		0		0.35	0.65	0.90	1.0			0.96	0.16	

		1	2	3	4	5	6	7	8	9	10	11	12
Individual	Initial Talent	Capital Period 1	Cap Period 2	Cap Period 3	Cap Period 4	Cap Period 5	Cap Period 6	Cap Period 7	Cap Period 8	Cap Period 9	Cap Period 10	Cap Period 11	Cap Period 12
1	0.7	1,500	1,500	750	375	188	188	563	141	47	23	23	6
2	0.68	500	125	42	21	7	21	21	21	63	63	21	7
3	0.43	500	125	31	10	10	21	21	10	10	10	10	10
4	0.65	500	500	250	250	250	500	500	1,500	1,500	1,500	6,000	6,000
5	0.55	500	500	250	83	83	83	42	42	42	14	14	14

FIGURE 8.93　Partial view of input data and results.

The evolution of capital for each individual in the following 79 periods is achieved by copying the following formula in the range D10:CD17009, parcial view in Figure 8.92:

IF(C10>=0.5;IF(RAND()<$B$4;IF(RAND()<$B10*(1+D$8*0.00125);LOOKUP (RAND();$D$5:$H$5;$E$3:$H$3)*C10;C10);C10/LOOKUP(RAND();$D $5:$H$5;$E$3:$H$3));0)

Cell J4 (Figure 8.93) shows the number of individuals who end up with at least one million monetary units:

=COUNTIF(CD10:CD17009;">=1000000")

The fortune accumulated by the luckiest of the thousands of individuals is shown in cell K4 (Figure 8.93) obtained by:

=MAX(CD10:CD17009)

Cell N4 (Figure 8.93) presents the number of individuals who ended up with less than 1 MU, who are qualified for bankruptcy:

COUNTIF(CD10:CD17009;"<=1")

In each execution, the maximum and minimum value of individual talent are determined:

Cell L5: =MAX(B10:B17009) and Cell M5: =MIN(B10:B17009)

Figures 8.94 and 8.95 provide the remaining expressions used for this experiment.

	CE	CF	CG	CH
1				
2	Successful Talented (C >500,000 & T >0.8) (Individuals)	Successful Untalented (C >500,000 & T <0.6) (Individuals)	Average Capital for Talented Population (MU)	
3	=COUNTIF(CE10:CE17009;1)	=COUNTIF(CF10:CF17009;1)	=AVERAGEIF(CG10:CG17009;">0")	
4				
5	Global Richness (MU)	Individual Average Net Worth (MU)		
6	=SUM(CD10:CD17009)	=CE6/17000		
7				
8				
9	Successful Talented? (C >500,000 & T >0.8) (0/1)	Successful Untalented? (C >500,000 & T <0.6) (0/1)	Final Wealth Amount of Talented Individual (T >0.8) (MU)	Talented & Millionaire?
10	=IF(AND(CD10>1000*$A$4;B10>0.8);1;0)	=IF(AND(CD10>1000*$A$4;B10<0.6);1;0)	=IF(B10>0.8; CD10;0)	=IF(AND(B10>=0.8;CD10>=10^6);1;0)
11	=IF(AND(CD11>1000*$A$4;B11>0.8);1;0)	=IF(AND(CD11>1000*$A$4;B11<0.6);1;0)	=IF(B11>0.8; CD11;0)	=IF(AND(B11>=0.8;CD11>=10^6);1;0)

FIGURE 8.94 Formulas to obtain various equity indicators.

CI	CJ	CK	CL
	Frequency Distribution of Individual Capital		
Richness Intervals (MU)	Individuals	%	Accumulated %
10	=FREQUENCY(CD10:CD10009;CI3:CI13)	=CL3	=(CJ3/$CJ$14)*100
=50	=FREQUENCY(CD10:CD10009;CI3:CI13)	=CL4-CL3	=CL3+(CJ4/$CJ$14)*100
100	=FREQUENCY(CD10:CD10009;CI3:CI13)	=CL5-CL4	=CL4+(CJ5/$CJ$14)*100
1000	=FREQUENCY(CD10:CD10009;CI3:CI13)	=CL6-CL5	=CL5+(CJ6/$CJ$14)*100
10000	=FREQUENCY(CD10:CD10009;CI3:CI13)	=CL7-CL6	=CL6+(CJ7/$CJ$14)*100
50000	=FREQUENCY(CD10:CD10009;CI3:CI13)	=CL8-CL7	=CL7+(CJ8/$CJ$14)*100
100000	=FREQUENCY(CD10:CD10009;CI3:CI13)	=CL9-CL8	=CL8+(CJ9/$CJ$14)*100
500000	=FREQUENCY(CD10:CD10009;CI3:CI13)	=CL10-CL9	=CL9+(CJ10/$CJ$14)*100
1000000	=FREQUENCY(CD10:CD10009;CI3:CI13)	=CL11-CL10	=CL10+(CJ11/$CJ$14)*100
5000000	=FREQUENCY(CD10:CD10009;CI3:CI13)	=CL12-CL11	=CL11+(CJ12/$CJ$14)*100
>5,000,000.00	=FREQUENCY(CD10:CD10009;CI3:CI13)	=CL13-CL12	=CL12+(CJ13/$CJ$14)*100
Total:	=SUM(CJ3:CJ13)	Individuals	

FIGURE 8.95 Expressions to determine the frequency.

In Excel the FREQUENCY function is a type array object. Therefore, the formula must be entered as an array formula =FREQUENCY(CD10:CD17009;CI3:CI13). **Ctrl Shift** and **Enter** keys must be pressed simultaneously to close with the symbol: }.

Randomness of Success: Summary of declared expressions.

Cell	Expression
E5	=D5+E4
F5	=E5+F4
G5	=F5+G4
H5	=G5+H4
B10	=MIN(ROUND(NORM.INV(RAND(); 0.6; 0.1); 2);1)
C10	=IF(RAND()<$B$4;IF(RAND()<$B10;LOOKUP(RAND();$D$5:$H$5;$E$3:$H$3))* $A$4; $A$4);$A$4/ LOOKUP(RAND();$D$5:$H$5;$E$3:$H$3))
C11	Copy the contents of cell C10 into column range C11 to C17009

(Continued)

D10	=IF(C10>=0.5;IF(RAND()<$B$4;IF(RAND()<$B10*(1+D $8*0.00125);LOOKUP(RAND();$D$5:$H$5;$E$3:$H$3)*C10;C10);C10/ LOOKUP(RAND(); $D$5:$H$5;$E$3:$H$3));0)
E10	Copy cell D10 from E10 to CD10. Then copy row range D10:CD10 in D11 to CD17009.
CE10	=IF(AND(CD10>1000*$A$4;B10>0.8);1;0)
CF10	= IF(AND(CD10>1000*$A$4;B10<0.6);1;0)
CG10	= IF(B10>0.8; CD10;0)
CH10	= IF(AND(B10>=0.8;CD10>=10^6);1;0)
CJ3:CJ13	FREQUENCY function is a type array object. Requires Ctrl Shift and Enter keys pressed simultaneously =FREQUENCY(CD10:CD17009;CI3:CI13)
CK3	=CL3
CK4	= CL4-CL3 Copy CK4 to range CK5:CK13
CL3	=(CJ3/$CJ$14)*100
CL4	CL3+(CJ4/$CJ$14)*100 Copy CL4 to range CL5:CL13
CE3	=COUNTIF(CE10:CE17009;1)
CF3	=COUNTIF(CF10:CF17009;1)
CG3	= AVERAGEIF(CG10:CG17009;">0")
CE6	= SUM(CD10:CD17009)
CF6	=CE6/17000
CG6	=SUM(CH10:CH17009)
J4	=COUNTIF(CD10:CD17009;">=1000000")
K4	= MAX(CD10:CD17009)
N4	=COUNTIF(CD10:CD17009;"<=1")
L5	=MAX(B10:B17009)
M5	= MIN(B10:B17009)

Proposed
Problems to Solve

9

9.1 PROBLEMS

---

1. A dealer's average car sales fit a Poisson distribution of an average of 5.2 cars/week. Simulate 10,000 weeks to estimate the probability of selling between two and three cars/week. Compare the estimated sample value (statistics) with the corresponding theoretical population value (parameter).

2. The failure time of a flat panel display is random Weibull with shape $\alpha = 1/3$ hours and scale $\beta = 200$. Simulate and estimate the probability that a display will fail before 2,000 hours.

3. On a production line, plastic bottles are automatically filled with liquid soap. The filled weight is normal random with average of 60 ounces and deviation of 1.25 ounces. Simulate the filling of 10,000 jars and estimate the percentage whose weight is between 59 and 61 ounces. Compare with theoretical value.

4. The lifetime of a LED lamp in hours has a gamma distribution with shape parameter $\alpha = 2$ and scale parameter $\beta = 25$. Simulate 100,000 lamps and estimate: (A) Probability that a lamp will last more than 50 hours. (B) Find the lamp average duration. (C) Use Data Analysis to construct the histogram of the times.

5. The following values correspond to the number of monthly stops at an automotive factory. Apply Chi square fit with a significance level of 0.05 to assess whether a Poisson model adequately represents the data.

24	11	11	16	13
7	9	12	16	11
11	13	18	14	8
15	10	11	15	10
9	15	8	14	8
16	8	17	12	6
12	8	7	7	7

DOI: 10.1201/9781032701554-9

6. The distances in centimeters between two perforations made in 60 metal sheets are shown below:

24.94	25.04	25.46	25.64	25.03
24.91	25.32	25.54	24.49	25.81
24.39	25.12	24.36	24.81	24.85
25.65	24.76	24.53	25.19	24.63
24.74	25.06	24.66	25.28	26.03
24.80	25.09	25.36	24.38	24.91
25.25	25.10	25.20	24.62	25.04
24.99	25.17	24.80	25.34	24.73
24.93	24.67	25.06	25.34	24.86
25.21	25.24	24.94	24.66	25.43
24.97	24.34	25.63	25.47	24.41
25.56	24.35	24.95	24.83	25.30

Determine if the distance values approximate a normal distribution with $\mu = 25$ cm and standard deviation $\sigma = 0.4$ at the 0.95 confidence level.

7. Customers at a bank branch line up in a queue to access two ATMs. The following values correspond to their interarrival times in minutes.

5.5	2.9	1.6	0.5	3.5
2.3	0.5	2.5	1.3	1.5
0.5	0.4	0.9	1.5	0.2
1.8	4.8	1	0.7	2.5
7.6	2.3	1.3	9.6	7.3
0.3	3.2	2.2	4	2.7
0.5	0.3	6.2	1.2	0.7
7.2	1.7	1	1.2	6.1
1	5.4	7.1	2.5	4.2
1.9	3.7	0.9	0.5	0.7

Service time is exponential, averaging 5 minutes. Fit an exponential distribution to the arrival data at 5% significance. Simulate for 8 hours a day and estimate the delay of customers in queue.

8. Consider the following downtime minutes in a manufacturing plant:

5.2	14.8	5.3	30.1	41.2
46.3	0.4	19	12.1	3.3
25.9	28.6	33.6	9.8	1.3
61.3	20	11.9	21.7	3.2
72.5	9.9	16.7	44.9	13.5
24.2	51.7	14.9	39.3	2.9
5.5	5.1	35.7	33.2	3.3
88.8	7	10.7	37.3	27.6
47	12.9	17.6	15.4	7.5
22.2	15.4	15	45.3	20.7

Use the chi-square test to assess whether a Weibull distribution fits this sample at the significance level $\alpha = 0.05$.

9. The duration time in days of 130 electrical accumulators for heavy machinery is shown in the table below:

845.4	996.7	948.2	830.8	898.5	954.8	998.6	899.9	929.6	1003.9
923.2	896.7	724.9	847.2	918.6	828.1	1133.3	923.6	995.9	872.9
794.4	696.8	970.5	887.7	769.8	966.7	865.8	977.1	899.3	923.1
871.2	874.0	922.7	1011.4	783.2	851.4	812.2	747.5	1032.4	1030.8
1007.1	950.3	925.5	842.4	904.5	873.0	876.3	957.5	911.9	968.6
1015.4	842.1	843.9	898.2	897.1	988.0	718.9	965.7	1097.8	860.8
900.2	802.4	821.3	891.9	804.7	800.9	954.3	937.0	948.2	880.2
928.0	1077.4	753.8	809.5	814.9	995.6	1035.3	938.8	935.7	935.3
835.2	795.4	1007.0	784.2	806.1	980.0	1108.4	949.9	893.0	928.4
1096.6	1027.1	1004.3	929.3	706.3	868.4	717.9	897.5	1036.7	926.9
870.3	729.7	787.7	1008.3	792.8	939.5	959.4	882.7	938.6	754.0
767.1	959.1	935.9	898.8	836.2	813.8	875.8	775.6	980.3	804.4
893.8	943.2	903.4	899.1	778.9	958.3	999.1	921.1	791.8	1052.1

Fit a probability distribution model and simulate 10,000 units to estimate the probability that an accumulator lasts at least 950 hours.

10. The time required to calculate the number of weekly hours worked by the employees of a textile company in minutes is:

1.2	4.1	5.9	1.2	4.5	1.5	3.3	5.3	1.7	10.0
2.1	3.8	0.7	8.7	0.0	6.4	8.3	3.9	10.5	0.6
1.3	7.9	2.6	1.1	6.4	3.2	6.5	5.4	2.8	4.5
6.2	3.0	2.1	8.9	1.2	0.3	2.0	2.0	10.4	1.3
2.1	3.9	2.4	4.6	2.2	1.2	6.5	2.5	1.3	2.9
2.4	0.3	2.4	2.4	4.0	6.8	1.7	4.2	0.7	0.7
1.5	3.8	9.5	5.5	0.7	1.9	9.0	2.3	8.5	22.4
0.2	1.8	4.5	10.3	7.4	9.3	9.0	3.1	6.7	0.7

Evaluate the fit of a gamma distribution with a significance level of 0.05. Simulate 10,000 calculations. Estimate the probability that the time to calculate the number of weekly hours for an employee is greater than 5 minutes.

11. Four dice marked D1, D2, D3, and D4 are rolled twice. Estimate probability of obtaining the same result in both launches. (A) Case 1: Each marked die repeats its number on the second roll. For example, Roll 1: (4 6 1 2) Roll 2: (4 6 1 2) is a favorable event. (B) Case 2: The brand of die does not matter. For example, Roll 1: (4 6 1 2) Roll 2: (1 6 4 2) is a favorable event.

12. Three balls are drawn successively and without replacement from an urn containing seven white balls and three black balls. Simulate 10,000 events to calculate the probability that the first two balls drawn are white and the third is black.

13. The height of plants two weeks after germination follows a normal distribution with a mean of 3.5 cm and a deviation of 0.5 cm. Simulate 10,000 plants and determine: (A) Probability that a plant exceeds 4 cm in height. (B) Probability that the height of the plant is between 2.5 and 3.5 cm.

14. The number of patients who request daily emergency care in a hospital has a Poisson average of 19.5 patients. The service collapses when the number of patients exceeds 25. Simulate 5,000 days and estimate the probability of collapse of the hospital's emergency department.

15. A tool box contains 10 large, 8 medium, and 6 small screws. Another box contains 8 nuts that fit with the large screws, 8 fit with the medium ones and 7 with the small ones. A bolt and a nut are chosen at random. Simulate 10,000 extractions and estimate the probability that the nut fits the screw.

16. The duration of a TV is modeled as a Weibull random variable with scale and shape parameters 12 years and 2.78, respectively. The manufacturer offers a 4-year warranty. For 10,000 TV, simulate and estimate with their respective 95% confidence intervals: (A) Percentage of times that the guarantee will have to be effective. (B) Average duration time without failure.

17. There are 15 people in a room. Their last names correspond to: 4 Smith, 5 Scott, and 6 Johnson. If three people are selected at random, simulate 10,000 times to estimate the probability that all three are surnamed Scott.

18. Two dice are thrown. Let the event A: Obtain an even value on dice 1. Let the event B: The sum of both dice is greater than 8. (A) Calculate the theoretical probability of $P(B/A)$. (B) For a confidence of 95% and an error of 1%, determine the simulation length to estimate the probability that the sum of both dice is greater than 8, given that an even value came up on the first dice.

19. According to car wash records, cars are serviced in three independent stages. Cleaning takes a normal time of average 20 minutes and deviation of 2.5. The washing follows an exponential average of 15 minutes. Drying is uniform between 8 and 14 minutes. The minimum cleaning or washing times are 10 minutes, respectively. Simulate the service of 1,000 cars to estimate the service duration.

20. To attract more customers, the owner of a fast food drive-through restaurant gives a free drink valued at $0.55 to the driver who waits more than 2.5 minutes to initiate service. Cars arrive at the single service window at the Poisson rate of 20 per hour. Attended by an employee, the service time is normal with a mean of 2 minutes/car, and a deviation of 0.5 minutes. The restaurant serves 24 hours a day. Simulate 10,000 cars and estimate: (A) Expected daily cost for free drinks. (B) Including waiting time, the probability of being served in less than 3 minutes. (C) The time that a customer spends on average in the system.

21. Mechanics working in a die-cutting plant must receive tools from a warehouse. An average of 10 mechanics arrive per hour looking for tools. Currently the warehouse is run by a warehouseman, who is paid $25 per hour and spends an average of 5 minutes delivering the tools required by a mechanic. The lost time cost of an idle mechanic waiting for tools is $6 per hour. The manager must decide whether or not to hire an assistant for the warehouseman, improving average service time by 1 minute. The helper would earn $15 per hour. Exponential times are assumed. (A) Simulate the arrival and attention of 50,000 mechanics. Determine: (A) Expected number of mechanics in the plant, without hiring the assistant. (B) Expected number of mechanics in the plant if the assistant is hired. (C) Expected cost per hour without assistant. (D) Expected cost per hour if the assistant is hired.

22. A system is evaluated in which trucks arrive at an unloading area attended by two crews. Each crew only serves one truck. There is enough waiting space. A truck is unloaded in exponential time of 12 minutes, and the trucks arrive at a Poisson ratio of eight trucks per hour. Simulate the

service of 10,000 trucks and estimate: (A) Expected time in the system. (B) Expected time in queue. (C) Average queue length. (D) Expected number of trucks in the system. (E) Waiting probability of a truck.

23. A game of chance consists of a roulette rotating mechanism that has 12 positions numbered from 1 to 12. When you operate a lever, a small ball begins to spin on the roulette wheel and is positioned on a winning number. In one turn, three people bet. The bettor only chooses one position. Bettors can agree on the same number. The amount bet per person is a uniform integer random variable between $50 and $100, but only in multiples of 5. If your number wins, you receive 10 times the value bet. (A) Determine the length of the simulation to estimate the expected utility of the game, with 95% confidence and precision of $1. (B) How much should the prize be for the game to be balanced?

24. A small-town fair betting game consists of a board with 1,000 holes. All the holes are sealed and inside contain a paper indicating the prize contained in the hole. A player bets $1 and selects a hole. The board has a $10 prize hole; five holes with $5 and 10 holes with a prize of $2. (A) Estimate the theoretical expected utility of the game. (B) Indicate if it is a fair game. (C) Use a pilot sample of 1,000 bets to determine the simulation length and estimate the expected utility of the game, with 95% confidence, and an error of one dollar.

25. A game of chance between a bettor and the house consists of the following: Two dice are thrown. If 7 is obtained the bettor wins. If the sum is less than 7, the house wins. If the sum is greater than 7, no one wins. On each throw, the bettor bets $10. If he wins, he receives $50. Simulate and estimate: (A) Expected value of the game, for a simulation length with 95% confidence and 3% error ($2.52). (B) Compare the estimated obtained value with the theoretical hope value of the game. (C) Probability of winning for the bettor.

26. A plane fighter will bomb a target measuring 150 × 150 meters by dropping three independent bombs aimed at the installation. The point of impact is normal random with an average of 150 meters and a deviation of 50 meters in each dimension. The target is only destroyed when it is hit by at least two bombs. Define a Cartesian framework and simulate 10,000 missions. Estimate the probability of destroying the target.

27. A land missile is launched on the installation defined in the previous exercise. The impact point is also normal with identical values. The missile only destroys if it hits within 25 meters of the target's geometric center. Simulate the launch of 10,000 shots and estimate the probability of destroying the target.

28. A hunter accurately shoots a deer with probability 0.35. Simulate 10,000 hunts to estimate the number of bullets to carry to have 95% success.

29. Four somewhat lying students did not attend the final exam. According to their story, they were in the same car and a tire burst. The teacher immediately isolates them and questions each one separately about which tire broke down. Since they did not expect it, each student randomly and independently indicates their answer. Simulate a length of ten thousand

events and estimate the probability that students will get away with your trick. Compare with theoretical probability.

30. An employee must travel 10 km by car daily to go from home to the office. It starts on a highway and then continues on a central avenue of the city. Every morning he accesses the highway, traveling 7 km. Sometimes he finds the road clear, and other times he suffers delays for various reasons. With a 0.65 probability it finds a clear path and its average speed for that section is 70 km/hour. With probability 0.35, it encounters delays and its average speed is only 25 km/hour. Then you enter an intersection controlled by a traffic light, programmed to last 150 seconds on red and 45 seconds on green. Finally, it travels an additional 3 km on the central avenue, traveling 35% of the time at 15 km/hour, 25% at 20 km/hour, 20% at 30 km/hour, and 20% of the time it encounters a traffic jam, which takes a normal random time of average 45 minutes and deviation of 10. Simulate 5,000 employee trips and estimate: (A) Average time to get to work. (B) Average time spent waiting at the traffic light. (C) How much time should you have to arrive on time with 75% confidence? (D) Sometimes there is an important meeting, and you want to be 99% confident that you will be on time. How much time should you have?

31. The Body Mass Index (BMI) is a quantitative method to assess the level of obesity of people. It is obtained from the quotient between weight and height squared. Assume the BMI for men aged 25 to 34 years as Lognormal, with $\mu = 3.215$ Kg/m^2, and $\sigma = 0.157$.

BMI values	Weight condition
≤18.0	Deficient
18.1–24.9	Normal
25.0–29.9	Overweight
30.0–34.9	Obese Type I
35.0–39.9	Obese Type II
≥39.9	Obese Type III (Frightening)

(A) Simulate a pilot sample of 1,000 men between 25 and 34 years old. Estimate the proportion of overweight men. (B) Use the thousand values to determine the simulation length to estimate the proportion of overweight men, with 95% confidence and an error of 5%.

32. In a factory, three types of chocolate bars are made and packaged according to their purity content A, B, and C. 25% are type A, 40% type B, and 35% type C. At the end of the day, the randomly six containers. Simulate 10,000 selections and determine the probability of choosing exactly two bars of each type.

33. A refinery uses two types of fuel to obtain jet gasoline. For unknown reasons, the amount of vapor pressure contributed by each fuel when mixed is a random variable whose distributions are indicated below:

Type I fuel

Vapor pressure	6	7	8	9	10	11	12
Probability	0.10	0.12	0.18	0.20	0.25	0.10	0.05

Type II fuel

Vapor pressure	8	9	10	11
Probability	0.30	0.25	0.30	0.15

Gasoline is obtained by mixing 350 liters of type I fuel with 650 liters of type II fuel, which are packaged in 1,000-liter barrels. If the vapor pressure of the resulting gasoline in a barrel is between 8.5 and 10.5, the barrel is accepted. Simulate the production of 10,000 barrels, and determine the probability of acceptance of one barrel.

34. The number of accidents per week in a factory follows a Poisson distribution with parameter $\lambda = 2$. Simulate 10,000 weeks to estimate: (A) Probability that in a week there will be an accident. (B) Probability that there will be two accidents in one week and another two in the following week. (C) Probability that there will be four or more accidents in two consecutive weeks.

35. The municipal government of a city plans the construction of a certain number of homes. The project contemplates three different models according to the size of the family group:

Type of housing	Number of children
A	0–1
B	2–3
C	3–4
D	5 or more

According to the National Census Office, the number of children of a family residing in the municipality approximates a Poisson variable with a mean of 3.5. In the municipality there are 5,000 families without housing. Simulate and estimate the number of homes by type to build.

36. Three machines A, B, and C make screws. In one hour, machine A produces 400 screws, machine B 350, and machine C 250. The probabilities that the machines produce defective screws are, respectively, 0.03, 0.04, and 0.05 for A, B, and C. At the end of production, all the screws are combined and one screw is chosen at random. Simulate the extraction of 10,000 screws to estimate the conditional probability that if is defective, it came from machine A, B, or C?

37. A producer plants 5 hectares of potatoes. According to their records, factors such as sales price, yield per hectare, and costs of fertilizers, irrigation, seeds, and labor are random according to the following probability distributions:

Price ($/ton)	200	300	400	500	600	700
Probability	0.10	0.15	0.35	0.25	0.10	0.05

Yield (ton/ha)	21	22	23	24	25	26
Probability	0.10	0.10	0.30	0.35	0.10	0.05

Cost ($/ton)	165	200	255	310
Probability	0.30	0.36	0.24	0.10

Simulate 10,000 crops and estimate the expected profit per hectare. Find the frequency distribution of the utility.

38. A company has four factories that produce the same item. The production of the factories is stored in the same warehouse. In a period, factories A, B, C, and D stored 20%, 40%, 25%, and 15%, respectively. The percentages of defectives per factory were 1%, 2%, 1.3%, and 0.5%. You take a product at random and it turns out to be defective. (A) Determine the length of the simulation to estimate the probability that the selected item comes from factory B, with a confidence level of 90% and an error of 1%. (B) Probability that the selected defective item comes from each of the factories.

39. A piece of equipment is made up of five circuits which fail with a certain frequency, causing the equipment to stop working while the circuits are replaced. The duration of a circuit is a uniform variable between 300 and 500 hours. We want to evaluate two replacement policies. Policy A consists of individually replacing the circuit that fails, in an exponential time of 2 hours on average. Policy B is to replace all five circuits when one or more circuits fail within 5 hours. The cost of the machine stopped is $100 per hour. The cost to replace a circuit is $20. Starting with five new circuits, simulate appropriately to evaluate both policies.

40. A football manufacturer produces three different models and offers a replacement guarantee if the product proves defective. In addition to the cost, and sales price per unit, the number of defective balls per model is a discrete Normal random variable, with average and deviation indicated below:

	Model A	Model B	Model C
Average number of defectives (in thousand balls):	16	12	8
Standard deviation:	3	3.5	2.2
Cost per ball ($):	9	12	17
Sale rice ($):	11	15	20

Monthly production is 1,000 balls of each model. Estimate: (A) Expected profit. (B) Expected cost for replacing defective balls.

41. A device contains two electrical components A and B. The life time of each component is distributed exponentially with an average of 5 and 10 years, respectively. The device works as long as both components work. Simulate 10,000 components and determine how long is the expected operating time.

42. On any given day a person is healthy or sick. If the person is fine one day, they have an 80% chance of being fine the next day. If you are sick on any day, the chance that you will still be sick the next day is 50%. It is assumed that a person's health only depends on their current condition. If today the person is healthy. Simulate 10,000 days to estimate the probability that the person will be sick three days later.

43. An electrical engineer must connect two resistors in parallel labeled as having 25 and 100 ohms of resistance each. The actual resistance of each device may vary randomly from its nominal values. R1 is N(100, 102) and R2 is N(25, 2.52). The resistance resulting from the assembly is:

$$R = \frac{R_1 * R_2}{R_1 + R_2}$$

The resistance of the circuit is required to be between [19, 21]. Simulate ten thousand circuits and determine the probability that the circuit assembly meets the specification.

44. Human Development Index of the member countries every year. The HDI is based on three indicators: Longevity (healthy and long life), educational level (possibility of acquiring knowledge), and standard of living (decent). Between 1970 and 2000, the HDI was collected for the different countries.

It was classified into three states: high if HDI >= 0.80, medium if 0.50 <= HDI <= 0.799, low if HDI < 0.50. For periods of 5 years, the following probability transition matrix was obtained:

	High	Medium	Low
High	0.885	0.115	0
Medium	0.084	0.896	0.02
Low	0	0.178	0.822

Estimate: (A) Probability that a medium HDI country will reach a high level after 20 years (0.396). (B) If there are 185 countries associated with the UN, estimate the number of countries expected to be positioned in each HDI level in the long term.

45. An insurance agency sells policies to five individuals of the same age. According to their records, the probability that an individual lives 20 more years is 0.6. (A) For a confidence level of 95% and error of 3%, determine the simulation length to estimate the probability that after 20 years at least three of the five insured people will be alive. (B) Estimate the probability that after 20 years at least three of the five individuals will live.

46. Auto Rental is a company that offers cars for rent. It has 500 units. In any given week a car can be in one of three states: Operational, undergoing minor maintenance, or major maintenance. Each car is inspected once a week. Within one week, 400 cars were determined to be operational, 80 needed minor repairs, and 20 required major repairs. In the following week, 350 of the cars that were in good condition were still operational, 40 needed minor repairs and 10 required major repairs. Of the 80 that required minor repairs, 50 were fine, 25 were still in minor repairs, and 5 required major repairs. Finally, of the 20 cars that required major repairs, 10 were operational, and 10 were still undergoing major repairs. This behavior is considered to be permanent (Markov chain) from week to week. Assume that a car in good condition produces normal random income of average $3,000 and deviation of $500. A car in minor repair produces expenses of $ $N(1,000;250)$ and in major repair $ $N(7,000;1,000)$. For a simulation length of 10,000 transitions (cars): (A) Simulate and estimate the expected weekly profit. (B) Annually, how many weeks is a car expected to spend in major maintenance?

47. A company establishes a pension fund for its employees by contributing 5% of each employee's salary and withholding 5% of each employee's salary. At the beginning, the company has a monthly payroll of 3 million MU. Every 6 months, the company must increase the salaries of all its

employees by 12%. Contributions to the fund are deposited at the beginning of each month in a special account, whose annual interest rate is a random variable whose behavior is a uniform random variable between 20% and 35%. Simulate 100 years to estimate the expected annual cumulative fund amount.

48. A clinic receives a weekly delivery of type AB-negative blood from the regional blood bank. Depending on availability, the quantity received in liters is random according to the following probability distribution:

Quantity (liters)	5	6	7	8	9	10
Probability	0.05	0.15	0.2	0.25	0.2	0.15

The number of patients per week requiring blood varies according to the following distribution:

Patients	0	1	2	3	4	5
Probability	0.10	0.15	0.20	0.25	0.15	0.15

The average requirement per patient is a uniform whole value between 3 and 6 liters. Assuming that the plasma is storable. Start with 2 liters in stock. If demand exceeds supply, the clinic is supported by the deficit shipment from a neighboring clinic. Simulate the reception and supply process for 10,000 weeks. Estimate: (A) Average plasma availability. (B) Average deficit.

49. The demand for sugar in a store is represented by a discrete exponential distribution of 100 kg/day. The store owner checks the inventory every 7 days, and places an order with the plant equal to the capacity of his warehouse minus the amount of sugar he has available at the time of the check. The plant supplies the order immediately. When it cannot meet the demand of its customers, the store incurs lost sales. The storage capacity of the tent is 700 kg. The ordering cost is $100 per order. The shortage cost is $6 per kg, and the inventory carrying cost is $1 per kg. (A) Determine the length of the simulation to estimate the average daily inventory cost, with a confidence level of 95% and error of 5%. Do 20 runs. (B) If you reviewed inventory every 5 days, would your cost improve?

50. The daily internal demand for parts to meet the repair needs of equipment used in a manufacturing process is a Poisson random variable with a mean of 5 units. At the end of each day, the number of pieces in stock is examined, and if there are less than 10 units, as long as there are no pending orders to receive, the supplier is ordered to send 10 units. It is known that the number of days required to receive the 20 units from the

supplier, counted from the order day, is a random variable with the following frequency distribution:

Days required by the supplier:	1	2	3	4
Observed frequency:	20	70	50	10

Assuming that operations begin with 15 pieces in stock, and with attention to pending orders due to lack of availability, simulate 10,000 days and determine: (A) Probability of ending any day with more than 4 units. (B) Probability of not having units available. (C) Average daily number of outstanding units. (D) Percentage of days that there is shortage. (E) How many units Q to order from the supplier to reduce the percentage of days with shortage pending units to 10%?

51. The weekly demand for a product follows a discrete triangular random variable, with minimum values of 60, maximum 140 and most probable 100 units. To optimize product stock management, the person responsible for maintaining inventory wants to determine the number of units to order from the supplier, in addition to the inventory replenishment level. The current inventory policy is as follows: at the end of each week the stock is reviewed. If it is less than 120 units, 400 units are ordered from the supplier. The supplier sells and supplies only batches of 5 units, and also guarantees that they will be available at the beginning of the following week. The cost of maintaining a unit in inventory is $20. If the inventory does not allow the entire demand to be met in a week, the deficit is partially satisfied with the quantity available, although an opportunity cost is incurred for each unit not served. The fixed cost of issuing an order to the supplier is $700. The purchase cost per unit is $800. The unit sales price is $950. At the beginning, there are 300 units. Simulate an appropriate simulation length and estimate: (A) Optimal inventory policy to follow. (B) Probability that the optimal policy does not allow all demand to be met in any week. (C) Expected order cost per year. (D) Probability that weekly demand is at least 85 units.

52. An investor wants to evaluate an investment project, for which he has the following information: The initial investment amount required is a triangular random variable, whose minimum value is $75,000 and maximum $400,000; although it is more likely $200,000. The expected income in the next 3 years are variables that are distributed according to the following information:

Year 1: Uniform between $20,000 and $60,000
Year 2: Normal with mean of $100,000 and deviation of $30,000

Year 3:

Income ($)	0	150.000	250.000	350.000	400.000
Probability	0.15	0.30	0.35	0.15	0.05

Evaluate the investment by estimating its Net Present Value. The interest rate for each year is a variable whose distribution is:

Annual interest rate	10%	13%	17%
Probability	0.30	0.45	0.25

Simulate 10,000 projects and estimate the Net Present Value of the investment.

53. Mr Jhonson is considering purchasing a bus to make trips from City1 to City2. Since it is a trip lasting more than 24 hours, he wants to offer the greatest comfort to his clients. Johnson can purchase a standard bus, although it may require some modifications. The specifications of a "standard" bus are 90 standard seats and 1 bathroom. He can add some "additional" ones. Between these:

   a. Place more comfortable armchairs in the first rows. To do this, it is necessary to use two standard seats for one armchair. According to design studies, the bus cannot have more than six seats and they must be placed in pairs.

   b. Place an additional bathroom. For design reasons, only a maximum of two bathrooms can be placed. For additional bathrooms, two standard stalls must be reduced.

According to records, the demand for standard seats follows a Normal distribution with a mean of 85 and variance of 25. The demand for armchairs follows a triangular distribution (0; 5; 7). The profit for each standard position is $200; and for each chair it is $500. Consider opportunity costs. An additional bathroom generates a profit of $450. Simulate and determine: (A) Simulation length to estimate the expected daily profit by optimal allocation of armchairs, standard seats, and bathrooms, with 95% confidence, and precision of $250. (B) Determine the probability that a traveler will not find a free chair. (C) Using a macro and 15 repetitions, estimate the expected utility. (D) What is the theoretical probability that the demand for standard positions on any day is at least 90 positions?

54. A car rental company evaluates the optimal car to buy for rent. The daily cost of a car is $250. The daily demand for cars requested for rental is distributed according to:

Car demand/day	2	3	4	5	6	7	8	9	10	11	12
Probability	0.03	0.02	0.08	0.12	0.20	0.15	0.15	0.10	0.07	0.05	0.03

If the daily income is $350/car/rented. The daily cost of an idle car is $250. Simulate 100,000 days and estimate the number of cars to be purchased.

55. The daily demand for a certain product is governed by a Binomial distribution with parameters n = 10 and p = 0.45. The delivery time in days is Poisson with an average of 3. The cost of keeping a unit in inventory is $2 per day. A missing unit costs $10. The cost of ordering from the supplier is $20 per order. We wish to compare two policies for maintaining inventory: (A) Order the necessary units every 3 days to bring the inventory to 20 items; and (B) Order 15 items when the inventory level is less than or equal to 10. It is assumed that the unsatisfied demanded units remain pending to be attended to when the items arrive. The sales price of each unit is $50 and its cost is $40. Start with 15 units in stock. Which of the two policies is more economical? It is assumed not to order again while there is a pending order to receive.

56. A tank can store up to 50,000 liters of water. Every 6 hours there is a requirement to discharge a random amount of water into a pipe network, according to a uniform distribution with parameters between 3,000 and 6,000 liters. However, it may deliver less or nothing as possible. Rainfall in the area occurs according to a Poisson distribution with an average of 12 rains per week. The amount of water that falls into the tank during each rain is a normal random variable with a mean of 9,000 liters and a deviation of 2,500 liters. The tank starts full. (A) Determine the simulation length to estimate the expected amount of water content at any time, with a confidence level of 99% and a maximum deviation of 500 liters. (B) Simulate the inlets and outlets of the liquid. Round the times at which events occur to integer values of the time of day. (C) Estimate the expected content of the tank at any time of day. (D) What is the probability that the tank at any time is empty?

9.2 ANSWERS TO PAIRED PROBLEMS

Below are the answers to the proposed even exercises. Does not include results from goodness-of-fit exercises. Chapter 6 contains numerous solved exercises on various data fit tests. Responses were obtained by averaging 20 repetitions. Sometimes each repetition corresponded to the simulation of 10,000 events. The alternatives to solve a simulation exercise are varied. Any result originating from simulation of random events is an approximation to the exact result. Although for simplicity it is not done here, all simulation results must be obtained after determining the sample size of the experiment or simulation length, in addition to being expressed with their respective confidence interval.

2. 0.884.

4. (A) 0.406. (B) 50 hours.

12. 0.175.

14. 0.0916.

16. (A) 0.0462. (B) 10 years, 80 days.

18. 0.0918.

20. (A) $81.2 per day. (B) 0.4626. (C) 4 minutes.

22. (A) 33.3 minutes. (B) 21.2 minutes. (C) 2.84 trucks. (D) 4.4 trucks. (E) 0.289.

24. E(U) = −0.945.

26. 0.1548.

28. 7 bullets (Figure 9.1).

Hunting	Rand Shooting	Shots Fired		Shots to Capture	Frequency	Accumulated % Hunts		
1	0.237	1		1	3525	35.25%		Hunts
	0.729			2	2236	57.61%		10,000
	0.750			3	1480	72.41%		Shoot
2	0.130	3		4	947	81.88%		
	0.849			5	614	88.02%		5
	0.559			6	436	92.38%		6
	0.516			7	265	95.03%		7

FIGURE 9.1 Hunter screen. Range A1:G8.

```
Sub Huntery()
 Randomize
 Range("A2.C1000").ClearContents
 p = Val(InputBox("¿Probability of Hitting a Shot? "))
 Hunt = Val(InputBox("How many hunts? "))
 Cells(2, "I") = Hunt
 Hit = "N"
 i = 1
 For k = 1 To Hunt
   Bullets = 1
   While Hit = "N"
     i = i + 1
     r = Rnd()
     Cells(i, "B") = r
     If r < p Then
       Hit = "S"
     Else
       Bullets = Bullets + 1
     End If
   Wend
   Cells(i, "A") = k
   Cells(i, "C") = Bullets
   Hit = "N"
   Next k
End Sub
```

30. (A) 27.3 minutes. (B) 0.96 minute. (C) 30 minutes. (D) 63.4 minutes.

32. 0.11.

34. (A) 0.864. (B) 0.073. (C) 0.565.

36. P(A/D) = 0.311. P(B/D) = 0.364. P(C/D) = 0.325.

38. P(A/D) = 0.139. P(B/D) = 0.574. P(C/D) = 0.233. P(D/D) = 0.054.

40. (A) $7,630. (B) $370.

42. 0.278.

44. High: 73 countries. Medium: 100 countries. Low: 12 countries.

46. (A) $1,962. (B) 2.9 weeks.

48. (A) 3.6 litters. (B) 4.1 litters.

50. (A) 0.628. (B) 0.232. (C) 9.45 units. (D) 40.27%. (E) 60 units (Figures 9.2–9.5).

Order Size Q	Reorder Level R		Poisson Average Demand	Units at Start		Supplier Delay	1	2	3	4
20	10		5	15			0.133	0.467	0.333	0.067
						0	0.133	0.600	0.933	1.000

Day	Is there a pending order?	Inventory at the beginning of the day	Poisson daily demand	Shortages + Demand	End of Day Inventory	Place an Order?	Delay in Arrival (days)	Day on which it arrives from the Supplier	Units Sold	Pending Units
1	N	15	3	3	12	N	--	--	3	0
2	S	12	7	7	5	S	2	5	7	0

FIGURE 9.2 Inventory system with attention to pending units.

Day	Is there a pending order?	Inventory at the beginning of the day	Poisson daily demand	Shortages + Demand
1	N	=E2	=INT(BINOM.INV($D$2/0.001;0.001;RAND()))	=D6
=A6+1	=IF(I6>A7; "S"; "N")	=IF(A7=I6; F6+$A$2; F6)	=INT(BINOM.INV($D$2/0.001; 0.001; RAND()))	=D7+K6

FIGURE 9.3 Formulas in the range A6:E10005.

End of Day Inventory	Place an Order?		Delay in Arrival (days)
=IF(D6>C6;0;C6-D6)	=IF(F6<=$B$2;"S";"N")		=IF(G6="S";LOOKUP(RAND();$G$3:$K$3;$H$1:$K$1);"--")
=IF(E7>C7; 0; C7-E7)	=IF(AND(F7<=$B$2; I6<=A7); "S";IF(I6="--"; "S"; "N"))		=IF(G7="S"; LOOKUP(RAND(); $G$3:$K$3; $H$1:$K$1); "--")

FIGURE 9.4 Formulas in the range F6:H10005

Day on which it arrives from the Supplier	Units Sold	Pending Units		P(Inventory = 0)
=IF(H6="--"; "--"; H6+1+A6)	=IF(C6>=D6; D6; C6)	=IF(C6<D6; D6-C6; 0)		=COUNTIF($C$6:$C$10005;0)/10000
=IF(H7="--"; I6; H7+1+A7)	=IF(C7>=E7; E7; C7)	=IF(C7<E7; E7-C7; 0)		

FIGURE 9.5 Formulas in the range K6:H10005.

52. $425,472.

54. 6 cars. A car rental.

56. (A) 79,000 events. (B) 16,500 liters. (C) 14,798 times. (D) 0.164 (Figures 9.6–9.11).

Input Data							Experiment Responses			
Deposit Capacity:	50,000	liters		Average	Deviation		A) Length	B) Content	Times the Tank is Empty	C) Probability.
	Min	Max	Input Rain:	9,000	2,500	liters	74,254	16,032	14,875	0.165
Volume to Download every 6	3,000	6,000	liters							
Average Rainfall (Poisson):	12	rains/week <===>	Time Between Rains (Exp):	14	hours/rain (Exp.)		Deposit Level Variance	215,795,557.5		

FIGURE 9.6 Input data and responses.

Time (hour)	Deposit Level at the Beginning (liters)	Random Next Rain R	Time Between Rains (hours)	Time the Next Rain R Will Fall (hour)	Time the Next Download D Will Occur (hour)	Random Estimate Download	Amount to Download (liters)	Amount to Enter the Deposit (liters)	Next Event R/D/ B (Both)	Deposit Amount Discharged (liters)	Deposit Level at End of Hour (liters)
0	50,000	0.2439	20	20	6	--	--	--	D	--	50,000
6	50,000	--	--	20	12	0.781	5,342	-	D	5,342	44,658

FIGURE 9.7 Partial view of evolution.

Time (hour)	Deposit Level at Beginning (liters)	Random Next Rain R	Time Between Rains (hours)	Time the Next Rain R Will Fall (hour)
0	=$C$2	=RAND()	=INT(-$G$6*LN(D9))+1	=+E9
=MIN(F9;G9)	=M9	=IF(OR(K9="R"; K9= "B"); RAND();"--")	=IF(OR(K9="R";K9="B"); INT(-$G$6*LN(D10))+1; "--")	=IF(OR(K9="R";K9="B"); F9+E10;F9)

FIGURE 9.8 Formulas in the range B9:F90008.

Time the Next Download D Will Occur (hour)	Random to Estimate Download	Amount to Download (liters)
6	--	--
=IF(OR(K9="D";K9="B"); G9+6;G9)	=IF(OR(K9="D";K9="B"); RAND(); "--")	=IF(OR(K9="D";K9="B"); INT($C$4+ ($D$4-$C$4)*H10); 0)

FIGURE 9.9 Formulas in the range G9:I90008.

Amount to Enter the Deposit (liters)	Next Event R/D/ B (Both)
--	=IF(F9>G9; "D";IF(F9=G9; "B"; "R"))
=IF(OR(K9="R";K9="B"); INT(NORM.INV(RAND();$G$3;$H$3)); 0)	=IF(F10>G10; "D";IF(F10=G10; "B"; "R"))

FIGURE 9.10 Formulas in the range J9:K90008.

Deposit Amount Discharged (liters)	Deposit Level at End of Hour (liters)
--	=C9
=IF(AND(I10>0; C10>I10); I10; IF(I10=0;0;C10))	=MIN($C$2;C10-L10+J10)

FIGURE 9.11 Formulas in the range L9:M90008.

Bibliography

Abramowitz, Milton, Stegun, Irene. *Handbook of Mathematical Functions with Formulas, Graphs and Mathematical Tables*. National Bureau of Standards.

Altiok, T., B. Melamed. *Simulation Modeling and Analysis with Arena*. Academic Press. Elsevier, 2007.

Banks, Jerry. *Handbook of Simulation: Principles, Methodology, Advances, Applications, and Practice*. Wiley-Interscience, 1998.

Bradshaw, J. "Software Agents. MIT Press, Cambridge, MA. Objects and agents compared." *Journal of Object Technology*, Vol. 1, No. 1, 41–53, 1997.

Burnham, Kenneth P., David R. Anderson. "Multimodel Inference: Understanding AIC and BIC in Model Selection." *Sociological Methods & Research*, Vol. 33, No. 2, November, 261–304, 2004.

Chung, Christopher. *Simulation Modeling Handbook: A Practical Approach*. CRC Press, 2004.

Cochkran, William G. "The X2 Test of Goodness of Fit." *The Annals of Mathematical Statistics*, Vol. 23, No. 3, 315–345, 1952.

Gottfried, Noether. "Note on the Kolmogorov Statistic in the Discrete Case." *Metrika*, Vol 7, February, 115–116, 1963.

Hawking, Stephen, Mlodinow Leonard. *The Grand Design*. Editorial Crítica, 2011.

Law Averill, Kelton W. *Simulation Modeling and Analysis*. McGraw-Hill, 1991.

Leemis, L. M. "Nonparametric Estimation of the Intensity Function for a Nonhomogeneous Poisson Process." *Management Science*, Vol. 37, No. 7, 886–900, 1991.

Montgomery, D., G. Runger. *Applied Statistics and Probability for Engineers. Limusa Wiley*. Limusa Wiley. 2002.

Nelson, Barry. *Stochastic Modeling Analysis & Simulation*. Dover Publications Inc, 1995.

Pooch, Udo, Wall James. *Discrete Event Simulation: A Practical Approach*. CRC Press, 2000.

Shenton, L., K. Bowman. "Remarks on Thom's Estimators for the Gamma Distribution." *Monthly Weather Review*, Vol. 98, No. 2, February, 154, 1970.

Stephens, M. A. "EDF Statistics for Goodness of Fit and Some Comparisons." *Journal American Statistics Association*. Vol. 69, 730–737, 1974.

Thom, Herbert C. "A Note on the Gamma Distribution." *Monthly Weather Review*, Vol. 86, 117–122, 1958.

Weibull, Walodi. "A Statistical Distribution Function of Wide Applicability." *Journal of Applied Mechanics*, Vol. 18, 293–297. 1939.

Ye, Zhi-Sheng, Nan Chen. "Maximum Likelihood-Like Estimators for the Gamma Distribution." *Department of Systems and Industrial Engineering*. National Singapore University, 2011.

Index

Printed in the United States
by Baker & Taylor Publisher Services

Printed in the United States
by Baker & Taylor Publisher Services